T0262979

Applications of Computational Intelligence Techniques in Communications

The book titled *Applications of Computational Intelligence Techniques in Communications* is a one-stop platform for researchers, academicians and people from industry to get a thorough understanding of the latest research in the field of communication engineering. Over the past decade, a manyfold increase in the use of computational intelligence techniques has been identified for obtaining the most optimal and timely solution to a problem. The various aspects covering the significant contribution of numerous computational intelligence techniques have been discussed in detail in this book.

In today's era of machine learning, the Internet of Things (IoT) is demanding an as high as possible data rate which has resulted in a tremendously increased speed of communication. To match up the pace, computational intelligence appears to be the most efficient and favourite tool. This book aims to cover the current technological advancements in the field of communication engineering and give a detailed prospect of computational intelligence to its readers.

This book will be a great support to people working in the field of IoT, machine learning, healthcare, optimization, filter design, 5G and beyond, signal processing etc. The chapters included here will expose its audience to various newly introduced and advanced computational intelligence techniques applicable in the communication domain. The readers will be introduced to multiple interdisciplinary areas of research in communication and will be inspired to work in collaboration with other professionals from both academia and industry.

Advances in Manufacturing, Design and Computational Intelligence Techniques

The book series editor is inviting edited, reference and text book proposal submission in the book series. The main objective of this book series is to provide researchers a platform to present state of the art innovations, research related to advanced materials applications, cutting edge manufacturing techniques, innovative design and computational intelligence methods used for solving nonlinear problems of engineering. The series includes a comprehensive range of topics and its application in engineering areas such as additive manufacturing, nanomanufacturing, micromachining, biodegradable composites, material synthesis and processing, energy materials, polymers and soft matter, nonlinear dynamics, dynamics of complex systems, MEMS, green and sustainable technologies, vibration control, AI in power station, analog-digital hybrid modulation, advancement in inverter technology, adaptive piezoelectric energy harvesting circuit, contactless energy transfer system, energy efficient motors, bioinformatics, computer aided inspection planning, hybrid electrical vehicle, autonomous vehicle, object identification, machine intelligence, deep learning, control-robotics-automation, knowledge based simulation, biomedical imaging, image processing and visualization. This book series compiled all aspects of manufacturing, design and computational intelligence techniques from fundamental principles to current advanced concepts.

https://www.routledge.com/Advances-in-Manufacturing-Design-and-Computational-Intelligence
-Techniques/book-series/CRCAIMDCIT

Series Editor: Dr. Ashwani Kumar

Sustainable Smart Manufacturing Processes in Industry 4.0 *edited by*
Ramesh Kumar, Arbind Prasad, Ashwani Kumar

Laser Based Technologies for Sustainable Manufacturing *edited by*
Avinash Kumar, Ashwani Kumar, Abhishek Kumar

Thermal Energy Systems: Design, Computational Techniques and Applications *edited by*
Ashwani Kumar, Varun Pratap Singh, Chandan Swaroop Meena, Nitesh Dutt

Additive Manufacturing in Industry 4.0: Methods, Techniques, Modeling, and Nano Aspects
edited by Vipin Kumar Sharma, Ashwani Kumar, Manoj Gupta, Vinod Kumar, Dinesh Kumar Sharma, Subodh Kumar Sharma

Advanced Materials for Biomedical Applications *edited by*
Ashwani Kumar, Yatika Gori, Avinash Kumar, Chandan Swaroop Meena, Nitesh Dutt

Applications of Computational Intelligence Techniques in Communications
Edited by Mridul Gupta, Pawan Kumar Verma, Rajesh Verma and Dharmendra Kr. Upadhyay

Applications of Computational Intelligence Techniques in Communications

Edited by
Mridul Gupta,
Pawan Kumar Verma,
Rajesh Verma and
Dharmendra Kr. Upadhyay

CRC Press
Taylor & Francis Group
Boca Raton London New York

CRC Press is an imprint of the
Taylor & Francis Group, an **informa** business

First edition published 2024
by CRC Press
6000 Broken Sound Parkway NW, Suite 300, Boca Raton, FL 33487-2742

and by CRC Press
4 Park Square, Milton Park, Abingdon, Oxon, OX14 4RN

CRC Press is an imprint of Taylor & Francis Group, LLC

ISBN: [9781032404196] (hbk)
ISBN: [9781032590356] (pbk)
ISBN: [9781003452645] (ebk)

DOI: 10.1201/9781003452645

Typeset in Times
by Deanta Global Publishing Services, Chennai, India

This book is dedicated to all engineers, researchers and academicians.

Contents

Preface

This book puts forward a special collection of contributions focused on important elements of Computational Intelligence (CI). In contrast to traditional computing, CI accommodates imprecise information, partial truth and uncertainty. The main components of CI that currently have frequent application in Communication Engineering are Signal Processing (SP), Machine Learning (ML) and Deep Learning. In CI, ML and DL are concerned with learning, while SP is concerned with precision and reasoning. This book primarily covers both the abovementioned key elements of CI, i.e., learning and analysis and modification of signals. All the contributions in this book have a direct relevance to ML from Optimization Techniques for Energy Optimization in Wireless Sensor Networks to advanced networks such as IoT-based Heterogeneous Nano Networks, Quantum Cryptography for Security in 5G and beyond 5G Communication Networks. The book also discusses Swarm Intelligence in Signal Processing. These are some of the Communication Engineering applications, where CI has excellent potential for use. Both beginner and expert readers should find this book a useful reference in the field of CI. The editors and the authors hope to have contributed to the field by paving the way for learning paradigms to solve real-world problems.

MATLAB® is a registered trademark of The MathWorks, Inc. For product information, please contact:

The MathWorks, Inc.
3 Apple Hill Drive
Natick, MA 01760-2098 USA
Tel: 508 647 7000
Fax: 508-647-7001
E-mail: info@mathworks.com
Web: www.mathworks.com

Acknowledgments

We express our gratitude to **CRC Press (Taylor & Francis Group)** and the editorial team for their suggestions and support during completion of this book. We are grateful to all contributors and reviewers for their illuminating views on each book chapter presented in the book *Applications of Computational Intelligence Techniques in Communications.*

Editors

Mridul Gupta obtained his B. Tech in Electronics and Communication Engineering in the year 2010 from Uttarakhand Technical University, Uttarakhand, India. He received his M. Tech from Motilal Nehru National Institute of Technology, Allahabad, India, in the year 2015 and his Ph.D. from Netaji Subhas Institute of Technology, University of Delhi, India, in 2021. He is currently working as an Assistant Professor with Graphic Era Deemed to Be University, Dehradun, India. He has published more than 20 research articles in various peer-reviewed international journals and conferences. He has also served as a reviewer for some reputed international journals like *IEEE Transactions on Microwave Theory and Techniques*; *Microwaves, Antennas and Wave Propagation*, IET; *Electronics Letters*, IET; *Microwave and Optical Technology Letters*, Wiley; *Chaos, Solitons & Fractals*, Elsevier. His research areas include optimization techniques, signal processing, fractional-order systems and microwave circuits.

Google Scholar page: https://scholar.google.co.in/citations?user=pcUWD8QA-AAAJ&hl=en

Pawan Kumar Verma obtained his B.E. in Electronics and Communication from Agra University, India and M. Tech. in VLSI Design from the C-DAC, Mohali, India, in 2005 and 2009, respectively. He has worked as a consultant for 2 years at Cadence Design Systems, Noida, India. He was a Visiting Research Scholar at the University of Waterloo, Canada, from April 2012 up to October 2012. He completed his Ph.D at the Motilal Nehru National Institute of Technology, Allahabad, India, in December 2016. Currently, he is working as an Assistant Professor at the National Institute of Technology, Jalandhar, India. His main research interests are M2M communications, Internet-of-Things, MANETs, VANETs, wireless networks and mobile computing.

Google Scholar Page: https://scholar.google.co.in/citations?user=Wsp4EOwAA AAJ&hl=en

Rajesh Verma is currently an Associate Professor in the Department of Electrical Engineering at King Khalid University, Abha, KSA. He obtained his B.E., M.E. and Ph.D. in Electronics and Communication Engineering from MNNIT, Prayagraj, India, in 1994, 2001 and 2011, respectively. He has more than 20 years of experience in teaching and administration at many reputed institutes in India including MNNIT-Prayagraj and many others. He also worked in the Telecom industry for 4 years in New Delhi, India. His research interests include computer networks, MAC protocols, wireless and mobile communication systems, sensor networks, peer-to-peer networks and M2M networks.

Google Scholar Page: https://scholar.google.co.in/citations?user=1pebRh4AAAAJ &hl=en

Dharmendra Kr. Upadhyay is a Professor in the Department of Electronics and Communication Engineering at Netaji Subhas University of Technology, New Delhi, India. Upadhyay has guided 6 Ph.D. theses and 36 M. Tech. theses. He has published more than 80 research papers in reputed international journals (41), Proceedings of International Conferences (42) and book chapters. He is a life member of ISTE. His research interests include Analog/Digital/Mixed Signal Processing and Antenna Designs.

Google Scholar Page: https://scholar.google.co.in/citations?user=fotlUiQAAAAJ &hl=en

Contributors

Abdelkader, Tamer
Ain Shams University, Cairo, Egypt

Abhi, Rayala Sai
Koneru Lakshmaiah Education
    Foundation, Andhra Pradesh, India

Agarwal, Meenakshi
Shri Vishwakarma Skill University,
    Palwal, Haryana, India

Alsabaan, Maazen
Department of Computer Engineering,
    College of Computer and
    Information Sciences, King Saud
    University, Riyadh, Saudi Arabia

Behera, Swabhiman
Department of Orthodontics, Saraswati
    Dental College, Lucknow, India

Gopina, Ram Koushik
Koneru Lakshmaiah Education
    Foundation, Andhra Pradesh, India

Goswami, Om P.
Amity University, Mumbai,
    Maharashtra, India

Gupta, Akash
Department of Electronics and
    Communication Engineering,
The LNM Institute of Information
    Technology, Jaipur, India

Gupta, Mridul
Graphic Era Deemed to be University,
    Dehradun, India

Jasper, Aakash
Motilal Nehru National Institute of
    Technology Allahabad, Prayagraj,
    Uttar Pradesh, India

Kandasamy, Gokulakannan
Centre for Rehabilitation, School of
    Health and Life Sciences (SHLS),
    Allied Health Professions, Teesside
    University, Middlesbrough, UK

Kande, Pavan Sai Gopal
Koneru Lakshmaiah Education
    Foundation, Andhra Pradesh, India

Khalifa, Tarek
American University of the Middle East

Kumar, Ashish
Agriculture Research Organization –
    Volcani Center, Israel

Kumar, Mahesh
Department of Computer Science and
    Engineering, Graphic Era Deemed to
    be University, Dehradun, India

Kumar, Mukul
Central Research and Incubation
    Center, Swami Vivekanand Subharti
    University, Meerut, India

Kumar, Naveen
University of Glasgow, Scotland

Kumar, Nawnit
Sri Guru Tegh Bahadur Institute of
    Management & IT, New Delhi, India

Kumar, Sachin
Department of Electronics and
    Communication Engineering,
    National Institute of Technology,
    Jalandhar, India

Man, Suresh
Department of Physiotherapy, Lovely
    Professional University, Punjab,
    India

Mittal, Deepak
Department of Mechanical
    Engineering, MIET, Meerut, Uttar
    Pradesh, India

Mishra, Shalabh K.
Bharati Vidyapeeth's College of
    Engineering, New Delhi,
    India

Pal, Raghavendra
Sardar Vallabhbhai National
    Institute of Technology, Surat,
    Gujarat, India

Paiva, Sara
ADiT-Lab, Instituto Politecnico de
    Viana do Castelo, Viana do Castelo,
    Portugal

Pathak, Shashwat
Atal Community Innovation Center,
    MIET Foundation, Meerut, Uttar
    Pradesh, India

Prakash, Arun
Motilal Nehru National Institute of
    Technology Allahabad, Prayagraj,
    Uttar Pradesh, India

Raghav, Sumit
Department of Physiotherapy, Lovely
    Professional University, Punjab,
    India

Shrivastava, Dhiraj
Department of Electronics and
    Communication Engineering,
The LNM Institute of Information
    Technology, Jaipur, India

Singh, Kamlesh
Department of Orthodontics, Saraswati
    Dental College, Lucknow, India

Singh, Kulwant
FlexMEMS Research Centre,
    Department of ECE, Manipal
    University Jaipur, Jaipur, India

Singh, Prabhdeep
Department of Computer Science and
    Engineering, Graphic Era Deemed to
    be University, Dehradun, India

Sharma, Abhay
Graphic Era Deemed to be University,
    Dehradun, India

Sharma, Shivani
Department of Electronics and
    Communication Engineering,
    National Institute of Technology,
    Jalandhar, India

Srivastava, Anirudh
Central Research and Incubation
    Center, Swami Vivekanand Subharti
    University, Meerut, India

Shikhola, Tushar
Delhi Metro Rail Corporation
    Limited, Netaji Subhas University
    of Technology (Formerly: Netaji
    Subhas Institute of Technology
    (University of Delhi))

Singh, Anshika
Subharti College of Physiotherapy,
    Swami Vivekanand Subharti
    University, Uttar Pradesh, India

Talluri, Sai Hemanth
Koneru Lakshmaiah Education
  Foundation, Andhra Pradesh, India

Taman, Omar Abd El-Rahman
Department of Computer Engineering
  and Software Systems, Ain Shams
  University, Cairo, Egypt

Tandon, Ragni
Department of Orthodontics, Saraswati
  Dental College, Lucknow, India

Tripathi, Rajeev
Motilal Nehru National Institute of
  Technology Allahabad, Prayagraj,
  Uttar Pradesh, India

Tripathi, Yogesh
Koneru Lakshmaiah Education
  Foundation, Andhra Pradesh, India

Verma, Bharat
Department of Electronics and
  Communication Engineering,
The LNM Institute of Information
  Technology, Jaipur, India

Verma, Pawan Kumar
Department of Electronics and
  Communication Engineering,
  National Institute of Technology,
  Jalandhar, India

Verma, Rajesh
Department of Electrical Engineering,
  King Khalid University, Saudi
  Arabia

1 An Introduction to Swarm Intelligence in Communication

Tushar Shikhola

1.1 INTRODUCTION

Swarm Intelligence (SI) is an important concept when it comes to Artificial Intelligence. In nature, Swarm-based Intelligence is seen in a diversified manner, e.g., colonies of ants, flocks of birds, schools of fish, herds of animals etc. These swarms possess a very unique phenomenon, i.e., a collective behaviour on a global scale, whereas individual swarms show restricted behaviour with the surrounding environment on a local scale. This collective global behaviour has been studied by researchers in an exhaustive manner in nature and simulated in the form of artificial swarms to solve large-scale optimization problems in the field of computational intelligence. Some of the well-established swarm intelligence-based optimization algorithms are Particle Swarm Optimization (PSO), Ant Colony Optimization (ACO), Grey Wolf Optimization (GWO), Whale Optimization Algorithm (WOA), Salp Swarm Algorithm etc. This chapter deals with the explanation of optimization techniques based on SI and their application in the field of communication technology focusing on Unmanned Aerial Vehicle (UAV) optimal placement problem and optimal path planning of UAV. Further, a brief summary of list of abbreviations used in the literature is mentioned in Table 1.1.

Considering the rapid pace of growth and recent advancements pertaining to communication technology, a Next-Generation Network has emerged that basically consists of different technologies, architectures etc. Optimization of such networks is of utmost importance.

Based on the application of game theory-based optimization algorithms and convex optimization, a recent innovation in wireless technology is the efficient application of an optimization technique known as Swarm Intelligence (SI).

SI is a part of the Artificial Intelligence (AI) field that derives its inspiration from the social behaviour of biological creatures such as flocks of birds, colonies of ants etc. SI paves the way for solving various complex and high-dimensional optimization challenges and helps to achieve promising results.

From the early 1980s, when the first generation of cellular network technology was developed, to date four generations of cellular networks have been introduced [1]. In

DOI: 10.1201/9781003452645-1

TABLE 1.1
List of Abbreviations

Abbreviations	Signifies	Abbreviations	Signifies	Keywords/ Acronyms	Signifies
5G	Fifth Generation	GA	Genetic Algorithm	NOMA	Non-Orthogonal Multiple Access
PSO	Particle Swarm Optimization	UAV	Unmanned Aerial Vehicle	GWO	Gray Wolf Optimization
ABC	Ant Bee Colony Optimization	BS	Base Station	SI	Swarm Intelligence
WSN	Wireless Sensor Network	WOA	Whale Optimization Algorithm	IoT	Inter of Things
ACO	Ant Colony Optimization	FA	Firefly Algorithm	NGN	Next-Generation Wireless Networks

the fifth generation (5G), it is considered that a large number of devices are interconnected to each other which shall lead to enhanced mobile broadband, massive usage of applications constructed with low latency connection and the Internet of Things (IoT). Considering various advantages of Next-Generation Networks, there are a number of challenges that shall be tackled to make the technology successful for its accessibility and penetration at a large scale. These challenges are highlighted below:

- Reduction in operational complexity and expenditure of network operators.
- User clustering and power control in Non-Orthogonal Multiple Access (NOMA) [2].
- Multi-access Edge Computing (MEC) systems resource allocation.
- Network slicing in virtualized wireless networks [3].
- Beamforming in systems with many antennas [4].

In the literature various techniques are available for solving the above problems, e.g., game theory-based approaches, convex optimization etc. However, the core concept for these algorithms to be applied is built on the notion that the objective function shall be convex in nature in order to determine the optimal solution to the given problem. But on the other hand, the feature of convexity is rarely applicable to the practical problem scenarios. This leads to limitations of these optimization algorithms. In such cases, heuristic algorithms find their applications in determining the optimal solution for the given non-convex optimization problems.

As a subset of AI, there has been widespread use of SI for solving complex optimization problems. SI-based optimization algorithms provide computationally light and robust solutions that are high quality in nature [5]. Various advantages of nature-inspired algorithms compared to conventional algorithms that are driven by gradient-based methodology are mentioned below:

- SI-based approach helps in determining the best possible solution to the problems in which internal mechanisms are very complex. In these problems, tuning the input parameters and noting the changes in the output help to determine the optimal solution.
- Using SI-based approaches, balance between exploration and exploitation are accomplished, improving the likelihood of discovering the best and most reliable solutions.
- In these approaches, information of gradient is not required as compared to other algorithms such as the Back Propagation algorithm. In this manner, issues such as trapping in the local optima and over-fitting are avoided in case of SI-based optimization approaches.
- SI-based algorithms are simple and easy for their implementation.

1.2   OVERVIEW OF SI

In this section, evolutionary algorithms are presented and a summary of well-known SI methods.

1.2.1   STOCHASTIC OPTIMIZATION AND BASICS OF SI

The term "*swarm intelligence*" was used in cellular robotic system which was empowered with Swarm Intelligence (SI). It refers to biological systems in which numerous individuals interact (popularly known as candidate solutions) to work together to accomplish a shared objective with the help of local interaction and with no centralized control. Consider a simple example found in nature, i.e., a colony of ants – there is a queen among the labourers, soldiers, etc. The main responsibility of the queen is to lay the eggs and other ants are responsible for developing the nest, defending the colony, collecting food etc. In the ant colony, the queen is not the commander and tasks are co-ordinated in an optimal sense to achieve designated goals. Each ant communicates locally with the help of pheromones and solves various complex problems and ensures the survival of their colony. In contrast to other types of algorithms such as evolutionary algorithms, event-based algorithms, SI-based algorithms or Meta-heuristics, SI-based algorithms are initialized with randomly generated solutions to the given optimization problem and with each iteration, and following the underlying methodology adopted by the algorithm, solution to the problem is determined when a satisfactory solution is reached complying to the defined conditions. However, deterministic algorithms do not have any random component. The typical structure of the stochastic optimization problem is shown in Figure 1.1

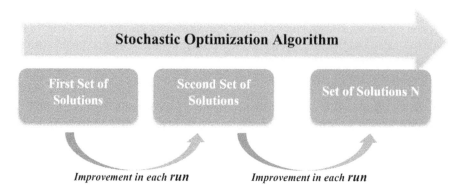

FIGURE 1.1 Typical framework of stochastic optimization algorithm

The set of solutions to the problem is called Candidate Solution as these solutions are the potential solution to the problem. Based on the number of Candidate Solutions, stochastic algorithms are classified as individual-based algorithms (one set of candidate solutions) and population-based algorithms (multiple candidate solutions) [6]. However, a natural curiosity arises on account of various pros and cons of deterministic and individual-based algorithms in comparison to population-based and stochastic algorithms. Table 1.2 describes in brief the various pros and cons of Population based Algorithms, Stochastic Algorithms and Deterministic Algorithms respectively.

Considering real-world issues/problems, boundaries or constraints must be taken into consideration while solving the given optimization problem or while evaluating the candidate solutions using SI-based techniques [6, 7]. Any candidate solution obtained that in case violates the constraints can be an optimal solution but shall not be practical and must be discarded. These solutions are popularly known as infeasible solutions. In order to handle infeasible solutions, there are various methods such as using appropriate penalty functions [8] which shall penalize those optimal solutions that violate the given constraints for a given problem and SI methods shall not favour these solutions as workable solutions.

1.2.2 Well-known SI Techniques

The field of SI falls within the umbrella of AI in which the algorithms are developed which tends to mimic the collaborative and collective phenomenon in nature in order to solve complex problems [9]. These methods are employed by a number of engineering disciplines to address optimization issues/problems.

A typical framework of SI-based methods is shown in Figure 1.1. Under this, a set of random solutions is the starting point for any optimization procedure. Each of the solutions in this set is essentially a potential, ideal solution to a certain problem. In order to evaluate the accuracy of each candidate solution, an objective function, often known as a fitness function is basically considered. With each iteration, this group of potential solutions gets better until a specific set of requirements is met.

TABLE 1.2
Pros and Cons of Various Types of Algorithms

S No.	Description	Pros	Cons
a)	Population-based algorithms	Avoid entrapment in local minima.	These algorithms are computationally expensive.
b)	Stochastic-based algorithm	Tends to avoid local optima problem due to presence of stochastic component.	Computationally expensive.
c)	Deterministic algorithms	i. Provides same solution in each run. ii. Computationally less expensive.	i. Tends to get trapped in local optima. ii. Predominantly requires gradient information of the optimization problem.
d)	Individual/Single solution- built-in algorithms	Tends to use single solution in each optimization stage and requires less processing power complexity compared to population-based techniques. Faster convergence.	Get entrapped in surrounding minima.

The primary characteristic of each algorithm, however, differs from the others due to the rules and principles that have been implemented to enhance the solution. Each potential solution, for instance, could be compared to an ant colony or a beehive. Finding the global minimum, or ideal solution to the actual problem, is the fundamental goal.

There are various reasons for using SI-based algorithms. These algorithms possess better capability to avoid the trap in local optima solutions, which is considered as one of the drawbacks of mathematics-based optimization algorithms. Further, SI-based algorithms are considered as black boxes, hence with the specified objective function, there is no requirement to compute the gradient information. Therefore, these algorithms are more practical for problem-solving. Finally, these methods use principles of nature which are coded as mathematical equations and thereby are able to obtain reasonable and practical global optimum solutions. However, the algorithms have the drawback of being computationally expensive in nature.

1.3 CATEGORIES OF SI

SI can be divided mostly into two categories:

- SI algorithms inspired by insects.
- SI algorithms based on animals.

In the next section some of the well-established and recent SI algorithms which are used in wireless and mobile networking are discussed briefly.

1.3.1 ARTIFICIAL BEE COLONY OPTIMIZATION

The Ant Bee Colony Optimization (ABC algorithm) method mimics the way bees naturally swarm to locate food. There is a population of solutions, just like in any other population-based algorithm. The three classes of individuals that make up this population are scouts, observers and worker bees. Each worker has a specific food source, and they all return to the nest together by performing the waggle dance. Before distributing food sources to workers, scout bees look for them. The waggle dance of other bees is used by observers to determine the food source [10].

1.3.2 PARTICLE SWARM OPTIMIZATION (PSO)

Kennedy and Eberhart introduced the Particle Swarm Optimization (PSO) algorithm, also known as PSO, in 1995, with the goal of locating the best solution for non-linear functions. This SI-based algorithm mimics the collective and individual intelligence of flocks of birds. An optimization problem is viewed as being in an n-dimensional space, where n is the number of parameters. Each bird in the flock stands for a partial solution. The swarm's collective best solution (global best) and the individual birds' best solutions are used by the birds to update their positions. In this way, PSO maintains the optimum answer to the specified optimization problem while searching around the most promising results produced by the particles [11].

1.3.3 ANT COLONY OPTIMIZATION

Ant Colony Optimization (ACO) was first presented in 1996 by (Dorigo & di Caro, 1999) [12]. It is a meta-heuristic technique that was developed for addressing combinatorial optimization problems and is inspired by nature. This algorithm takes inspiration from stigmergy, which is the process of communicating by modifying the swarm environment.

In order to send specific signals to other ants, ants communicate by leaving pheromone deposits on the ground or other objects. In a colony, they employ several pheromones for a variety of purposes. For instance, to draw other ants for transportation, they mark various routes from their colony to a food source. The shortest channel is automatically selected utilising this straightforward communication since the longer path faces a longer amount of time for vaporization before re-depositing the pheromone. The goal of the original ACO problem, which was presented as a graph with each ant representing a tour, was to find the shortest path. Distance and pheromone are the two metrics used in this technique. Each ant updates its pathways using both of them. A tour will be formed after multiple iterations and the update of both matrices, through which all ants will pass. This will be regarded as the most accurate evaluation of the problem.

1.3.4 FIREFLY ALGORITHM

A meta-heuristic algorithm known as the "firefly algorithm" (FA) tends to mirror the natural behaviour of fire flies during mating. A population of various random solutions is produced by this algorithm. The primary equation for this algorithm's location updating is as follows:

$$x_i = x_i + \beta_0 + e^{\gamma r_{ij}} (x_j - x_i) + \alpha \varepsilon_i$$

In the above-mentioned equation, ε_i is a random number in the range [0 1], x_i shows the position of the ith solution, γ represents the coefficient of light absorption, β_0 is the appeal/attractiveness from a distance, $r_{ij} = 0$ and α is the randomization parameter.

1.3.5 GREY WOLF OPTIMIZATION ALGORITHM

The social structure of wolves and their hunting habits are simulated by Grey Wolf Optimization (GWO) algorithm. This algorithm starts with a set of randomly generated solutions and iteratively develops them. The three best solutions found throughout the optimization process are utilized to update the positions of the other solutions. These top answers illustrate how well alpha, beta and delta wolves can hunt. The motivation comes from the fact that the prey's position is typically better known to the alpha, beta and delta wolves. One of the most popular modern metaheuristics, the GWO algorithm has a variety of applications that show its effectiveness in resolving practical problems [13].

1.3.6 WHALE OPTIMIZATION ALGORITHM

The Whale Optimization Algorithm (WOA) takes its cues from humpback whales' bubble net foraging technique. Producing and constricting a net of bubbles to draw the school of fish to the surface is a clever technique for collecting them. A group of whales are regarded as potential solutions for a particular optimization problem, much like GWO. Two mechanisms – **circling prey and spiral movement** – update their locations. The first strategy is comparable to GWO, however the second updates the position using a spiral equation [14].

1.4 APPLICATION OF SWARM INTELLIGENCE IN IOT NETWORKS

SI has recently demonstrated its ability to offer adaptable and wise solutions for the path placement and path planning design of UAV in NGN.

1.4.1 UAV PLACEMENT

SI algorithms have demonstrated their enormous potential in enhancing UAV-related communication designs, such as UAV positioning, by taking into account the idea of self-organized behaviour in nature [15]. Presented a framework for three-dimensional (3D) UAV installation that allows for wireless coverage of high-rise buildings' interior environments during earthquakes. A PSO-based algorithm is used to locate the exact location of the UAV under the random distributions of users in order to minimize the transmit energy needed to cover the building. A 3D UAV placement for indoor users in small coverage areas is explored in Sawalmeh et al. [16], and the model created there can be employed in wireless convergence for big crowded events. In this study, the UAV altitude placement is designed to achieve the best height for both single and multiple UAV scenarios while decreasing transmission power. In the study "Another UAV Placement Solution for Very Crowded Events," [15] by utilizing the ground-based BSs, UAVs can offer wireless coverage for both interior and outdoor customers. Both Air-to-Ground and Outdoor-to-Indoor route loss models are used to optimize the data rates of indoor and outdoor users under a UAV transmit power budget in order to specify the best 3D UAV deployment. In the meanwhile, the study in [17] focuses on the issue of QoS awareness and 3D UAV placement in ad hoc wireless networks. A PSO optimization technique is used to solve the joint formulation of each station's transmit power and number of aerial BSs. According to the simulation results, the joint scheme can significantly improve performance and offer an energy-efficient solution for 3D UAV placement of BSs in an ad hoc network.

1.4.2   PATH PLANNING OF UAV

Duan et al. [18] suggested simulating the UAV path-planning models using a bio-inspired optimization learning environment (Figure 1.2). In this work, a student-learning process is carried out for supporting the UAV path planner located at the ground control station using SI-based techniques such as PSO, ABC and ACO. The

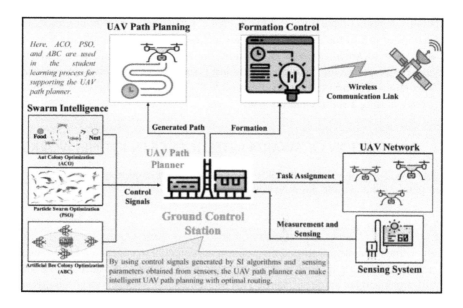

FIGURE 1.2   SI optimization-based UAV path-planning system [18]

UAV path planner may create intelligent UAV route planning with optimal routing for power reduction by combining control signals produced by SI algorithms with measurement and sensing parameters collected from sensors.

As compared to directed-search approaches, SI-inspired algorithms exhibit better advantages in terms of strong robustness, simplicity of implementation and high flexibility in a variety of UAV path-planning scenarios, according to the statistical analytic results. In 2018, Causa et al. [19] developed a more effective ACO-based UAV path-planning architecture that takes into consideration obstacles in a controlled space with several radars. The travelling salesman problem is resolved by creating an ant simulation of the salesman route in cities in order to create a feasible path for UAVs. In order to tackle the described path-planning problem, a multi-colony ACO algorithm is developed, which performs better than the traditional ACO scheme. Moreover, the problem of path planning for UAV swarms is also resolved [20] such that each UAV is thought of as an intelligent agent that searches for its objectives in reconnaissance scenarios by updating its position and speed. To produce the UAVs' best pathways for quick convergence and low latency target recognition, a distributed PSO model is used [21]. Additionally, in order to reduce energy consumption and enhance the UAV's ability to reject disturbances, a set-based PSO algorithm with adaptive weights was employed to apply the best path-planning strategy.

REFERENCES

1. Q. V. Pham et al., "A Survey of Multi-access Edge Computing in 5G and Beyond: Fundamentals, Technology Integration, and State-of-the-Art," IEEE Access, vol. 8, pp. 116974–117017, 2020, doi: 10.1109/ACCESS.2020.3001277.
2. L. Dai, B. Wang, Z. Ding, Z. Wang, S. Chen, and L. Hanzo, "A Survey of Non-Orthogonal Multiple Access for 5G," IEEE Communications Surveys and Tutorials, vol. 20, no. 3, pp. 2294–2323, 2018, doi: 10.1109/COMST.2018.2835558.
3. Y. K. Tun, N. H. Tran, D. T. Ngo, S. R. Pandey, Z. Han, and C. S. Hong, "Wireless Network Slicing: Generalized Kelly Mechanism-Based Resource Allocation," IEEE Journal on Selected Areas in Communications, vol. 37, no. 8, pp. 1794–1807, 2019, doi: 10.1109/JSAC.2019.2927100.
4. T. M. Pham, R. Farrell, and L.-N. Tran, "Revisiting the MIMO Capacity With per-Antenna Power Constraint: Fixed-Point Iteration and Alternating Optimization," IEEE Transactions on Wireless Communications, vol. 18, no. 1, pp. 388–401, 2019, doi: 10.1109/TWC.2018.2880436.
5. Q.-V. Pham, T. Huynh-The, M. Alazab, J. Zhao, and W.-J. Hwang, "Sum-Rate Maximization for UAV-Assisted Visible Light Communications Using NOMA: Swarm Intelligence Meets Machine Learning," IEEE Internet of Things Journal, vol. 7, no. 10, pp. 10375–10387, 2020, doi: 10.1109/JIOT.2020.2988930.
6. H. Faris, I. Aljarah, M. A. Al-Betar, and S. Mirjalili, "Grey Wolf Optimizer: A Review of Recent Variants and Applications," Neural Computing and Applications, vol. 30, no. 2, pp. 413–435, 2018, doi: 10.1007/s00521-017-3272-5.
7. A. H. Gandomi and X.-S. Yang, "Evolutionary Boundary Constraint Handling Scheme," Neural Computing and Applications, vol. 21, no. 6, pp. 1449–1462, 2012, doi: 10.1007/s00521-012-1069-0.

8. Ö. Yeniay, "Penalty Function Methods for Constrained Optimization with Genetic Algorithms," *Mathematical and Computational Applications*, vol. 10, no. 1, pp. 45–56, 2005, doi: 10.3390/mca10010045.

9. X.-S. Yang, "Chapter 1 - Introduction to Algorithms," in *Nature-Inspired Optimization Algorithms*, X.-S. Yang, Ed. Oxford: Elsevier, pp. 1–21, 2014, doi: 10.1016/B978-0-12-416743-8.00001-4.

10. D. Karaboga and B. Basturk, "A Powerful and Efficient Algorithm for Numerical Function Optimization: Artificial Bee Colony (ABC) Algorithm," *Journal of Global Optimization*, vol. 39, no. 3, pp. 459–471, 2007, doi: 10.1007/s10898-007-9149-x.

11. J. Kennedy and R. Eberhart, "Particle Swarm Optimization," in *Proceedings of the ICNN'95 - International Conference on Neural Networks*, vol. 4, pp. 1942–1948, 1995, doi: 10.1109/ICNN.1995.488968.

12. M. Dorigo and G. di Caro, "Ant Colony Optimization: A New Meta-Heuristic," in *Proceedings of the 1999 Congress on Evolutionary Computation-CEC99 (Cat. No. 99TH8406)*, vol. 2, pp. 1470–1477, 1999, doi: 10.1109/CEC.1999.782657.

13. S. Mirjalili, S. M. Mirjalili, and A. Lewis, "Grey Wolf Optimizer," *Advances in Engineering Software*, vol. 69, pp. 46–61, March 2014, doi: 10.1016/J.ADVENGSOFT.2013.12.007.

14. S. Mirjalili and A. Lewis, "The Whale Optimization Algorithm," *Advances in Engineering Software*, vol. 95, pp. 51–67, May 2016, doi: 10.1016/J.ADVENGSOFT.2016.01.008.

15. H. Shakhatreh, A. Khreishah, A. Alsarhan, I. Khalil, A. Sawalmeh, and N. S. Othman, "Efficient 3D Placement of a UAV Using Particle Swarm Optimization," in *8th International Conference on Information and Communication Systems (ICICS)*, vol. 2017, pp. 258–263, 2017, doi: 10.1109/IACS.2017.7921981.

16. A. Sawalmeh, N. S. Othman, and H. Shakhatreh, "Efficient Deployment of Multi-UAVs in Massively Crowded Events," *Sensors*, vol. 18, no. 11, 2018, doi: 10.3390/s18113640.

17. B. Perabathini, K. Tummuri, A. Agrawal, and V. S. Varma, "Efficient 3D Placement of UAVs with QoS Assurance in Ad Hoc Wireless Networks," in *28th International Conference on Computer Communication and Networks (ICCCN)*, vol. 2019, pp. 1–6, 2019. doi: 10.1109/ICCCN.2019.8846947.

18. H. Duan, P. Li, Y. Shi, X. Zhang, and C. Sun, "Interactive Learning Environment for Bio-Inspired Optimization Algorithms for UAV Path Planning," *IEEE Transactions on Education*, vol. 58, no. 4, pp. 276–281, 2015, doi: 10.1109/TE.2015.2402196.

19. F. Causa, G. Fasano, and M. Grassi, "Multi-UAV Path Planning for Autonomous Missions in Mixed GNSS Coverage Scenarios," *Sensors*, vol. 18, no. 12, 2018, doi: 10.3390/s18124188.

20. Y. Wang, P. Bai, X. Liang, W. Wang, J. Zhang, and Q. Fu, "Reconnaissance Mission Conducted by UAV Swarms Based on Distributed PSO Path Planning Algorithms," *IEEE Access*, vol. 7, pp. 105086–105099, 2019, doi: 10.1109/ACCESS.2019.2932008.

21. R.-J. Wai and A. S. Prasetia, "Adaptive Neural Network Control and Optimal Path Planning of UAV Surveillance System With Energy Consumption Prediction," *IEEE Access*, vol. 7, pp. 126137–126153, 2019, doi: 10.1109/ACCESS.2019.2938273.

2 Applications of Computational Intelligence in Societal Welfare

Meenakshi Agarwal

2.1 INTRODUCTION

Intelligence is the capabilities that allow an individual to think, learn and acquire knowledge and behave accordingly. Intelligence may be categorized as human intelligence and Computational Intelligence (CI). As far as humans are concerned, on the one hand, they have the abilities of experiential learning, adaptability to new situations, understanding, handling of abstract concepts and to use knowledge to control the systems. On the other hand, humans are also sensitive and have feelings of joy, excitement, boredom, tiredness, sadness etc. Therefore, in doing repetitive tasks, humans become tired due to the monotony. The other limitations applicable to humans are limited computation, limited communication and limited time [1]. To solve the abovementioned problems, the idea of developing a human-like intelligence developed which is known as CI [2]. It is currently one of the research areas with the fastest growth, attracting researchers from across the world. It has aided a wide range of social applications, including banking, finance, healthcare, image identification, cloud computing, advanced farming, energy management, cyber security, smart robotic weaponry for military use, disaster management, business intelligence and many others [3–8].

This chapter discusses how CI can be utilized for the welfare of human beings. For example, if one observes the field of agriculture, it is one of the complicated sectors where growing different crops throughout the course of a year requires knowledge, effort and technology [9]. In addition to the kind of soil and fertilizers, the monsoon season has an impact on Indian agriculture. Moreover, it is crucial to routinely monitor harvest and post-harvest analyses as well as crop disease and damage detection [10]. In this, machine intelligence can be utilized to analyze agricultural activities closely in order to reduce human efforts. Another sector where CI is highly recommended to overcome the limitations of humans is in the field of medical science. In several medical diagnostic systems, artificial and human intelligence are

DOI: 10.1201/9781003452645-2

integrated to diagnose various disorders with high accuracy [11]. Similarly, in order to reduce food wastage and improve food safety, data analysis and machine intelligence can be used [12].

In this chapter, a number of CI-based applications are discussed. This exploratory study provides a broad perspective for several applications of intelligent monitoring across a range of fields.

2.2   METHODOLOGIES USED FOR COMPUTATIONAL INTELLIGENCE

To outperform human intelligence, human-like models of machines are being developed. Humans have put in efforts to build technological models of themselves [13]. The purpose is not fulfilled without CI. There are several methods that assist developers to design machines. A few of them are as follows.

2.2.1   NEURAL NETWORK

The Neural Network is a network or a circuit composed of either natural or synthetic neurons. The first is referred to as a Biological Neural Network (BNN), and the second is referred to as an Artificial Neural Network (ANN). The applications in which artificial networks can be trained using a dataset include adaptive control, predictive modelling and other uses. Within networks, which can draw conclusions from a complicated and seemingly unconnected set of information, self-learning resulting from experience can take place. This is the rationale behind why CI experts are involved in the creation of ANN [13].

2.2.2   FUZZY LOGIC

Fuzzy Logic (FL) is a technique for making fast decisions. This strategy is comparable to human decision-making options between YES and NO. It differs from Boolean Logic, which solely distinguishes between true and false. FL was invented by Lotfi Zadeh. He perceived that, in contrast to computers, humans have a wider range between YES and NO options as depicted in Figure 2.1.

| Definitely Yes |
| Probably Yes |
| Not Certain |
| Probably No |
| Definitely No |

FIGURE 2.1   Levels for fuzzy logic: Incorporated

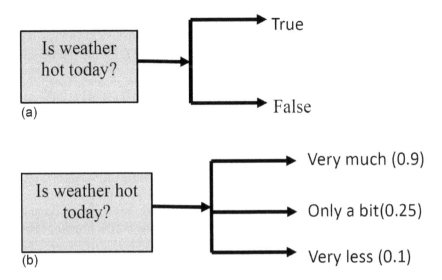

FIGURE 2.2 (a) (b) Fuzzy Logic: Incorporated

Therefore, FL represents all intermediate possibilities between TRUE and FALSE logics. For example, if we are discussing today's weather, Boolean Logic will produce a definite output as TRUE or FALSE, which is equivalent to a human being's YES or NO. On the flip side, FL will put together three to four levels of possibilities between YES and NO as depicted in Figure 2.2 (a) and (b).

Furthermore, FL is used in numerous applications such as natural language processing, speed control, traffic control, intelligent highway systems and various intensive applications in Artificial Intelligence (AI) [14].

2.2.3 EVOLUTIONARY COMPUTATION

A branch of AI known as "evolutionary computation" is a series of global optimization algorithms that draws inspiration from AI. It is widely used to handle issues with too many variables for conventional methods and difficult optimization problems [15]. Evolutionary algorithms such as genetic algorithms, evolutionary programming, genetic programming and swarm intelligence models like ant colony optimization or particle swarm optimization are used by computers that do evolutionary computing. Also, it is used to solve a variety of scheduling and DNA analysis issues [16].

2.2.4 PROBABILISTIC METHODS

Probabilistic methods are non-constructive methods used in mathematics. It shows that if the objects are chosen randomly from a specified class, the resulting probability of the prescribed kind is strictly greater than zero. It offers clear guidelines for making decisions and substantive assurances on the quality of those choices [17]. The recommendations are predicated on the idea that normative knowledge represents an important abstraction of genuine human experience. That is comparable to its counterpart in factual knowledge, which may be encoded and manipulated to generate helpful recommendations. On the other hand, decisions about the likelihood of events are made using probabilities, while decisions about the action's and results' attractiveness are measured using utilities [18].

2.2.5 LEARNING THEORY

One of the key CI techniques that searches for a form of "reasoning" similar to that of humans is learning theory. Learning, according to psychology, is the process of acquiring, enhancing or changing knowledge, skills, beliefs and worldviews by combining emotional, contextual and cognitive influences and experiences [19]. Learning theories then assist in comprehending how these impacts and experiences are processed, and it then aids in understanding how predictions are produced based on prior experience.

2.3 APPLICATIONS OF CI FOR THE WELFARE OF HUMAN BEINGS

It is understood that CI possesses a characteristic of intelligence which is attributed to human beings. Nowadays, several products and machines claim to be "intelligent" as they are able to reason and carry out decision-making similar to humans [20]. Therefore, CI is being used to develop products that reduce manual tasks and make human life more comfortable. Due to this reason, CI is being used in a plethora of applications [21]. A few examples are as follows.

2.3.1 SMART AGRICUTURE

Agriculture is one of the complex fields where skills, technology and hard work are involved to grow various crops during a year [9]. Moreover, Indian agriculture is closely related to the type of soil and fertilizers as well as monsoon season. Application of CI has been evident in the agriculture sector including harvest and post-harvest analysis along with disease and damage detection of crops which are important to be monitored regularly [10]. Therefore, extensive research has been made to apply CI at several stages of agriculture including the following.

- *Soil management*: Soil management is an integral part of agriculture. Farmers prepare the soil in the first step of farming to grow crops [21].
- Fuzzy logic, Management-Oriented Modelling (MOM) and decision support systems can be applied [21].
- *Sowing seeds*: For seeds to be sowed, this stage demands maintaining the precise spacing between seeds and depth [22].
- *Addition of fertilizers*: Maintaining soil fertility with the addition of fertilizers of the right kind is crucial for producing crops that are nourishing and healthy [23].

- *Growth monitoring*: Continuous monitoring of crops is essential to identify healthy plants and remove unhealthy ones [21].
- *Irrigation*: This action serves in maintaining humidity levels and tracking soil moisture. If done improperly, under- or overwatering the soil can stunt plant growth and result in ruined crops [22].
- *Weed management*: Unwanted plants known as weeds commonly appear next to crops or at field edges. Weed impacts crop quality, yield and production costs directly or indirectly. Hence, weed control is essential to cultivate healthy crops at a lower cost [22].
- *Harvesting*: Harvesting is the process of removing ripe crops from the field. It is a labour-intensive operation since it calls for a number of workers. Moreover, post-harvest management includes cleaning, sorting, packing, refrigeration etc. in this stage [21, 22].
- *Crop storage*: At this stage of agriculture, the products are preserved safely throughout this post-harvest phase of the industry to ensure food security. Crop transportation and packaging are also included [22].

2.3.2 HEALTHCARE

Health monitoring is one of the primary sectors which is closely related to human beings. The main target of health-related applications is to analyze the results of clinical tests made on patients and the expected outcomes [23]. Additionally, CI is also useful at several stages such as diagnosis of disease, development of drug, personalized medicine, patient monitoring and care. AI technology differs from traditional technologies for its ability to collect data, clean and process data and produce a well-defined output to the end-user. These complex tasks can be potentially done by CI. However, results solely depend upon applied inputs [24].

2.3.3 FOOD QUALITY MANAGEMENT

Life cannot be thought about without food. Therefore, maintaining the quality and safety of food becomes important for one's life. Computational systems can assist to reduce food wastage and improve food safety [25]. It is also helpful to remove the chemicals and pesticides deposited at the outer layers of fruits and vegetables. The systems gather data from smart sensors and analyze it to forecast food quality and shelf life in order to track the state of the food. Particularly, monitoring is difficult due to a number of factors, such as the supply and demand for energy and the dependability of the sensors [26].

2.3.4 CYBER SECURITY

When dealing with cyber security issues in complex systems like the Internet of Things (IoT), Cyber-Physical Systems (CPS), etc., CI is crucial. The exponential expansion of devices has resulted in a considerable increase in security difficulties, including those related to automatic system vulnerabilities, data security concerns, malware detection, risk to public and personal safety and management of stored data. As a result, the management of CI approaches is made simple by their applicability in a wide range of applications, including cyber security, privacy, technologies for cyber defence, techniques for intrusion detection and security of data in IoT [27].

2.4   CONCLUSIONS

With the advent of technology, CI, also known as human-like intelligence has improved the daily life of human, significantly. Further, humans are sensitive and have limited computational ability and time. On the flip side, a machine does not feel monotonous in doing repetitive tasks for a long time. Therefore, CI has drawn the attention of researchers and scientists from all over the world. For example, the amalgamation of CI with healthcare technologies has greatly transformed the healthcare field. Several smart sensors-based devices are developed which are monitoring the health of a person at his home. He need not visit the hospitals frequently.

Similar to healthcare, a wide variety of fields such as modern business, agriculture, healthcare monitoring, development of medicines, automobiles, data handling, processing of data and safety of data etc. are being transformed by CI and reducing their dependency on manual tasks.

As a result, CI has made contributions to a wide range of societal applications, such as banking, finance, medical science, image recognition, agriculture, cloud computing, advanced farming, energy management, cyber security, smart robotic weapons for military applications, disaster management, business intelligence and many others.

REFERENCES

1. Griffiths, Thomas L. Understanding human intelligence through human limitations. *Trends in Cognitive Sciences*, 2020, 24.11: 873–883.
2. Diamant, Emanuel. Computational intelligence: Are you crazy? Since when has intelligence become computational? In *2016 IEEE Symposium Series on Computational Intelligence (SSCI)*. IEEE, 2016; 1–4.
3. Liu, Yue; Ghandar, Adam; Theodoropoulos, Georgios. A metaheuristic strategy for feature selection problems: Application to credit risk evaluation in emerging markets. In *2019 IEEE Conference on Computational Intelligence for Financial Engineering & Economics (CIFEr)*. IEEE, 2019; 1–7.
4. Shen, Tianyu, et al. Parallel medical imaging: An ACP-based approach for intelligent medical image recognition with small samples. In *2021 IEEE 1st International Conference on Digital Twins and Parallel Intelligence (DTPI)*. IEEE, 2021; 226–229.
5. Sulistyo, Susanto B.; Woo, Wai Lok; Dlay, Satnam Singh. Computational intelligent color normalization for wheat plant images to support precision farming. In *2016 Eighth International Conference on Advanced Computational Intelligence (ICACI)*. IEEE, 2016; 130–135.
6. Asim, Muhammad, et al. A review on computational intelligence techniques in cloud and edge computing. *IEEE Transactions on Emerging Topics in Computational Intelligence*, 2020, 4.6: 742–763.

7. Xue, Jin-Lin, et al. An agricultural robot for multipurpose operations in a greenhouse. *The DEStech Transactions on Engineering and Technology Research*, Res. 2017: 122–131.

8. Wu, Jui-Yu. Computational intelligence-based intelligent business intelligence system: concept and framework. In *2010 Second International Conference on Computer and Network Technology*. IEEE, 2010; 334–338.

9. Nicolas, Chollet; Naila, Bouchemal; Amar, Ramdane-Cherif. TinyML smart sensor for energy saving in internet of things precision agriculture platform. In *2022 Thirteenth International Conference on Ubiquitous and Future Networks (ICUFN)*. IEEE, 2022; 256–259.

10. Lešić, Vinko, et al. Rapid plant development modelling system for predictive agriculture based on artificial intelligence. In *2021 16th International Conference on Telecommunications (ConTEL)*. IEEE, 2021; 173–180.

11. Farkhadov, Mais; Eliseev, Aleksander; Petukhova, Nina. Explained artificial intelligence helps to integrate artificial and human intelligence into medical diagnostic systems: Analytical review of publications. In *2020 IEEE 14th International Conference on Application of Information and Communication Technologies (AICT)*. IEEE, 2020; 1–4.

12. Henrichs, Elia; Krupitzer, Christian. Towards adaptive, real-time monitoring of food quality using smart sensors. In *2022 IEEE International Conference on Autonomic Computing and Self-Organizing Systems Companion (ACSOS-C)*. IEEE, 2022; 70–71.

13. Reznik, Leonid. General principles and purposes of computational intelligence. *Systems Science and Cybernetics-Volume III*, 2009, 198.

14. Zhang, Yan-Qing; Lin, Tsau Young. Computational web intelligence (CWI): Synergy of computational intelligence and web technology. In *2002 IEEE World Congress on Computational Intelligence. 2002 IEEE International Conference on Fuzzy Systems. FUZZ-IEEE'02. Proceedings (Cat. No. 02CH37291)*. IEEE, 2002; 1104–1107.

15. Kubota, Naoyuki. Computational intelligence for cognitive robotics. In *2019 IEEE International Conference on Cybernetics and Computational Intelligence (CyberneticsCom)*. IEEE, 2019; 1.

16. Piuri, Vincenzo; Scotti, Fabio. *Implementations of Computational Intelligence Techniques*. Institute of Electrical and Electronics Engineers, 2007.

17. Paul, S.; Mitra, A.; Rajulu, K. G. From probabilistic computing approach to probabilistic rough set for solving problem related to uncertainty under machine learning. In *2015 IEEE International Conference on Computational Intelligence and Computing Research (ICCIC)*, Madurai, India, 2015; 1–6.

18. Pearl, Judea. *Probabilistic Reasoning in Intelligent Systems: Networks of Plausible Inference*. Morgan kaufmann, 1988.

19. Wang, Yingxu. On abstract intelligence and brain informatics: Mapping the cognitive functions onto the neural architectures. In *2012 IEEE 11th International Conference on Cognitive Informatics and Cognitive Computing*. IEEE, 2012; 5–6.

20. https://en.wikipedia.org/wiki/Computational_intelligence.

21. Eli-Chukwu, Ngozi Clara. Applications of artificial intelligence in agriculture: A review. *Engineering, Technology & Applied Science Research*, 2019, 9.4: 4377–4383.

22. Vadlamudi, Siddhartha. How artificial intelligence improves agricultural productivity and sustainability: A global thematic analysis. *Asia Pacific Journal of Energy and Environment*, 2019, 6.2: 91–100.

23. Davi, Caio Cesar Medeiros; Silveira, Denis Silva; de Lima Neto, Fernando Buarque. A framework using computational intelligence techniques for decision support systems in medicine. *IEEE Latin America Transactions*, 2014, 12.2: 205–211.

24. Cui, Jie. Computer-aided analysis and treatment of air-conditioning system failure in clean operating room of hospital. In *2022 Second International Conference on Artificial Intelligence and Smart Energy (ICAIS)*. IEEE, 2022; 945–948.

25. Henrichs, Elia; Krupitzer, Christian. Towards adaptive, real-time monitoring of food quality using smart sensors. In *2022 IEEE International Conference on Autonomic Computing and Self-Organizing Systems Companion (ACSOS-C)*. IEEE, 2022; 70–71.

26. Henrichs, Elia. Enhancing the smart, digitized food supply chain through self-learning and self-adaptive systems. In *2021 IEEE International Conference on Autonomic Computing and Self-Organizing Systems Companion (ACSOS-C)*. IEEE, 2021; 304–306.

27. Zhao, Shanshan, et al. Computational intelligence enabled cybersecurity for the internet of things. *IEEE Transactions on Emerging Topics in Computational Intelligence*, 2020, 4.5: 666–674.

3 Computational Intelligence in Cell-Level Healthcare Application

Mahesh Kumar, Prabhdeep Singh,
Omar Abd El-Rahman Taman,
Ashish Kumar, and Kulwant Singh

3.1 INTRODUCTION AND ORIGINS

Microfluidic devices for healthcare diagnostics have been developed over the last two decades. Diagnostics with microfluidic devices is accomplished at the cell or enzyme level and has the potential for early-stage disease information. These systems utilized for diagnostics with microfluidic devices include syringe pumps for medium infusion, data acquisition systems, microscopic instruments and signal-conditioning circuits with filters [1–3]. Microfluidic sciences deal with the physiochemical attributes of the cells or enzymes. The cell-level early-stage diagnostic is more attractive and relevant for the diagnostic of fatal diseases such as cancer as its impact on the human body propagates over time and a piece of early information is crucial for analyzing the drug to be used. The unique power of microfluidic devices is that they can rapidly analyze and quantitatively measure many simultaneous parameters of biological cells and can sort them as per utilization [4]. The main attribute of these systems is that they can sense and accurately analyze the dielectric properties and behaviour of cells in a low conductive buffer medium. This allows using the microfluidic device in cell caging, cell separation, cell focusing, cell trapping and cell levitation. The efforts of researchers are primarily going on in miniaturizing lab setups, automation, efficient sensor data transmission, efficient signal conditioning, electrode design optimization and channel dimension optimization. Microfluidic systems mainly evolved from five fields of science and engineering which are electronics, biology, mechanical, chemistry and computer science. These systems with the combination of science and engineering are mainly utilized in four diagnostic applications which are cell analysis, enzyme analysis, organ-level diagnostics and whole blood analysis with potential markets in hospitals, pathological labs and individual households as shown in Figure 3.1.

Biology is the most important field in microfluidics for certain important medical reasons as multiple biological cells have distinct properties and are required to be

DOI: 10.1201/9781003452645-3

19

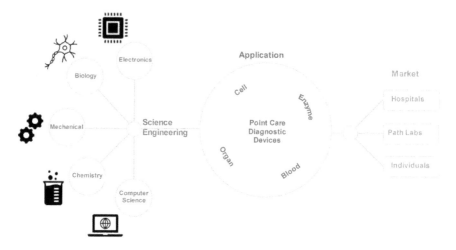

FIGURE 3.1 Microfluidic systems evolution and its potential application with market utilization

separated with high accuracy. The cell possesses different elements such as mitochondrion, lysosome, Golgi vesicles, nucleus, microtubes, ribosome and plasma (or cell membrane). A cell can be mainly considered to have two main elements: cytoplasm and membrane with reference to its dielectric properties which are important with respect to cell motion manipulation. Many companies are nowadays developing microfluidic cell separation systems for cell counting and for the purpose of analysis. The first time a cell counter was used was by the inventor Wallace Coulter at the time of World War II [5]. Coulter first applied this in the counting of plankton particles that always caused large echoes. Because the coulter counter is highly versatile and has high accuracy it is widely used in different areas of research including medical and life sciences. When the Coulter initially exhibited the Coulter counter it was able to detect and count up to 6000 cells per second of blood. Coulter counter can also have efficiency in the measurement of impedance and conductance when a single particle or cell passes.

Current-stage microfluidic devices or systems have the potential to sort nearly ten thousand cells per second. However, sorting efficiency depends upon environmental conditions such as temperature, atmospheric pressure and the type of cells to separate. At the initial level the basic requirements of microfluidic devices were always for detection but after the invention of the coulter counter the importance of cell separation and counting came into research. The device used for the detection of the cell is known as a flow cytometer. The term flow cytometer came from flow (fluidic suspension), cyto (concerning biological cells) and meter (a measuring instrument). This field of analyzing cell behaviour in a fluidic medium is known as flow cytometry.

Impedance flow cytometers are mainly used to measure the change of impedance across the microelectrodes of the microfluidic device and give output in terms of voltage and current [6]. When these electrodes are excited with alternating potential at specific frequencies the cells experience a force due to the dipole moment inside the cell. Microelectrodes used in flow cytometers are fabricated in the microfluidic channel of the device and are used to check the dielectric behaviour of the cell/particle suspended in the fluidic medium.

3.2 CURRENT STAGE OF CELL-LEVEL MICROFLUIDIC RESEARCH

Biological cell plays a major role in human health and can provide sufficient information for early-stage healthcare diagnostics for many fatal diseases when cells get damaged [7]. Some diseases are critical that can become incurable over time so detection of such diseases at the cell level is very much desired [8]. Before studying the cell behaviour, the biological sample is processed and the cells are separated. The devices which are used for the separation of biological cells are microfluidic devices and are generally termed lab-on-chip devices. There are various cell motion manipulation techniques utilized by researchers to isolate multiple cells depending on their physical attributes. The most common techniques for this purpose include fluorescence-activated cell sorting (FACS) technology [9], magnetically activated cell sorting (MACS) technology [10], centrifugation [11], inertial microfluidics [12], hydrophones [13], acoustic technology [14], filtration methods [15] and dielectrophoresis (DEP) technology [16].

Thousands of cells can be sorted within seconds using FACS technology but require complex equipment, costly setups and personalized personnel for operation and maintenance [9, 17]. MACS technology also serves as a good method with less complex setup but cells which are stimulating to magnetic field only can be influenced [10, 18]. Both the FACS and MACS technology need labelling of biological cells before processing which is quite cumbersome for rare pathogenic cell identification. Centrifugation and inertial microfluidics both manipulate the motion of the cells based on their size and density, so cells which are similar in physical properties are hard to separate [11, 12]. Hydrophoresis and acoustic technology utilize pressure manipulation in the suspending fluid in a microchannel so multiple cell sorting is quite tough to achieve [12, 13]. Filtration methods are also utilized by some of the research groups but microfilters (MFs) suffer from high sample loss and this method also separates cells based on size only [15, 19]. To avoid such issues and achieve separation of biological cells based on both physical and intrinsic electrical properties, DEP technology is more attractive [16]. The DEP technique works based on dipole movement inside cells and in the fluidic medium while under the influence of a non-uniform electric field [20]. The advantages of the DEP technology include no switching or moving elements in the device, label-free approach and miniaturized setup.

3.3 PRINCIPLE OF DIELECTROPHORESIS

Dielectrophoresis was explained for the first time by the researcher H.A. Pohl in 1958 [20], where the effects of non-uniform electric fields on micro-entities such as cells and microparticles suspended in dielectric fluidic medium were demonstrated. The electric field was applied externally in such a way that some parts of the device had distribution electric field lines and other parts with concentrated electric field lines. When the micro-entity was placed in between distributed and concentrated fields its direction of movement was based on its dielectric behaviour. More polarizable cells than the medium moved in concentrated electric field and micro-entities with lower polarizability than medium moved towards distributed electric field. There were two forces on the micro-entity such that one force (the DEP) actuated it in a lateral direction as discussed above and another force (the ER) influenced the cells with rotational force in either a clockwise or anticlockwise direction.

3.3.1 DIELECTROPHORESIS FOR HOMOGENEOUS MICROPARTICLES

Homogenous microparticles such as polystyrene particles are used in the medical field for replicating the behaviour of biological cells for preliminary use. These microparticles come in various sizes from 2 μm to 10 μm with specific conductivity and dielectric permittivity. When these particles are suspended in a fluid having a non-uniform electric field the dipole present in the particles tries to align with electric field lines passing through it. This causes a force which drags the particle towards a stronger or weaker electric field depending upon polarizability of the particle and medium. The generalized mathematical expression of this dielectrophoretic force for homogeneous microparticles is given by Equation (1).

$$F_{\mathrm{DEP}} = 2\pi\varepsilon_{\mathrm{m}}R^3 \operatorname{Re}\left[f_{\mathrm{cm}}\right]\nabla\left|E_{\mathrm{rms}}\right|^2 \tag{1}$$

In the DEP drag force, the real part of the Clausius Mossotti factor is the main analyzing parameter given by $\operatorname{Re}[f_{\mathrm{cm}}]$. When the cells are required to migrate in a concentrated electric field then the value for frequency should be maintained such that the values for $\operatorname{Re}[f_{\mathrm{cm}}]$ for that entity should be positive and when cells are expected to move towards a distributed electric field then the operating frequency should be selected such that it gives the $\operatorname{Re}[f_{\mathrm{cm}}]$ values below zero or negative. When the $\operatorname{Re}[f_{\mathrm{cm}}]$ values are positive then the induced dielectrophoretic force is said to be positive dielectrophoresis (p-DEP), and when the $\operatorname{Re}[f_{\mathrm{cm}}]$ values are negative then the induced force is known as negative dielectrophoresis (n-DEP). The mathematical expression for the Clausius Mossotti factor is given by Equation (2).

$$f_{\mathrm{cm}} = \left(\frac{\varepsilon_{\mathrm{c}}^* - \varepsilon_{\mathrm{m}}^*}{\varepsilon_{\mathrm{c}}^* + 2\varepsilon_{\mathrm{m}}^*}\right) \tag{2}$$

The Clausius Mossotti factor mainly depends upon two complex permitivities: (1) $\varepsilon_{\mathrm{m}}^*$: complex dielectric permittivity of the suspending fluidic medium, and (2) $\varepsilon_{\mathrm{c}}^*$: complex dielectric permittivity of the micro-entity (cell or particles). These complex permitivities are function of frequency and the mathematical expression for both permitivities are given by equations (3) and (4).

$$\varepsilon_{\mathrm{c}}^* = \varepsilon_{\mathrm{c}} - i\frac{\sigma_{\mathrm{c}}}{\omega} \tag{3}$$

$$\varepsilon_m^* = \varepsilon_m - i\frac{\sigma_m}{\omega} \tag{4}$$

In the above equations the σ_c and σ_m are electrical conductivities of cell and fluidic medium, respectively. Using equations (2), (3) and (4), the Clausius Mossotti factor (f_{cm}) can be further expressed as equations (5) and (6).

$$\mathrm{Re}\left[f_{cm}\right] = \frac{\left[(\varepsilon_c - \varepsilon_m)(\varepsilon_c + 2\varepsilon_m) + \dfrac{1}{\omega^2}(\sigma_c - \sigma_m)(\sigma_c + 2\sigma_m)\right]}{\left[(\varepsilon_c + 2\varepsilon_m)^2 + \dfrac{1}{\omega^2}(\sigma_c + 2\sigma_m)^2\right]} \tag{5}$$

$$\mathrm{Im}\left[f_{cm}\right] = \frac{\dfrac{1}{\omega^2}\left[(\sigma_c - \sigma_m)(\varepsilon_c + 2\varepsilon_m) - (\varepsilon_c - \varepsilon_m)(\sigma_c + 2\sigma_m)\right]}{\left[(\varepsilon_c + 2\varepsilon_m)^2 + \dfrac{1}{\omega^2}(\sigma_c + 2\sigma_m)^2\right]} \tag{6}$$

The above equations are used to analyse the DEP forces and electro-rotational (ER) forces on the micro-entity. The DEP force and ER force with respect to $\mathrm{Re}[f_{cm}]$ and $\mathrm{Im}[f_{cm}]$ are shown in Figure 3.2.

While analysing the frequency response of $\mathrm{Re}[f_{cm}]$, at some frequencies the values for $\mathrm{Re}[f_{cm}]$ is observed as positive and for other frequencies the $\mathrm{Re}[fcm]$

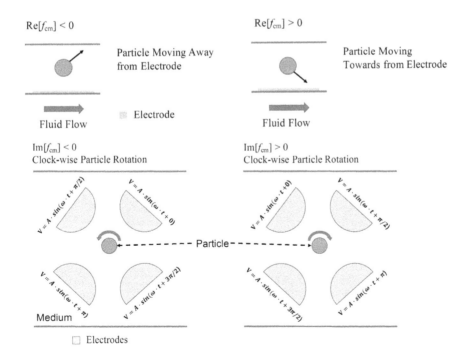

FIGURE 3.2 Electro-rotation of a biological cell in quadrupole electrodes

values are negative. The frequency where the $Re[f_{cm}]$ values change its polarity is known as cross-over frequency which can be calculated by equating $Re[fcm] = 0$ in equation (5).

3.3.2 DIELECTROPHORESIS FOR NUCLEATED BIOLOGICAL CELLS

The biological cell mainly contains cytoplasm and plasma as discussed earlier and possess different conductivity and dielectric permittivity. The difference in these parameters of different biological cells can be used for separation purpose. The behaviour of biological cell suspended in fluidic medium can be understood with two cases similar to microparticles. When cells have less complex dielectric permittivity then the medium gets repulsive force from electrode. When cells have more complex dielectric permittivity then the medium gets attractive force from electrode. The DEP forces acting on a nucleated cell is shown in Figure 3.3.

3.4   COMPUTATIONAL ANALYSIS FOR CELL-LEVEL APPLICATION

Forces on biological cells can be utilized for their separation, based on their dielectric response in the given medium as discussed in the previous section. Some of the cells produce only one type of dielectric response like in the yeast cells which only produces positive DEP response in DI water but some cells which are nucleated shell type produces both negative and positive dielectric response at different frequencies cells. The dielectric behaviour is observed in terms of $Re[f_{cm}]$ values calculated on different frequency values. The different parameters for calculation of dielectric response for various cells and polystyrene particles are represented in Table 3.1. The dielectric response shown in Figure 3.4 is calculated with the help of MATLAB code provided in appendix. Figure 3.4 shows the results for dielectric response for polystyrene particles, RBCs, PLTs and yeast cells when suspended in medium DI water. Here the computed response gives the information about the frequency at which system should be operated to achieve desired direction of movement of the cells. For example, when RBCs are desired to move away from the electrode then the $Re[f_{cm}]$ value should be negative for the operating frequency. With the calculated results in Figure 3.4 it is clear that polystyrene particle possesses negative DEP effect

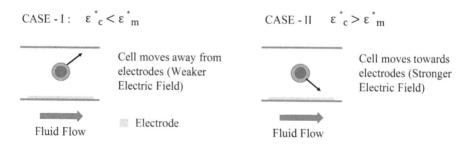

CASE - I : $\varepsilon^{*}_{c} < \varepsilon^{*}_{m}$ CASE - II $\varepsilon^{*}_{c} > \varepsilon^{*}_{m}$

Cell moves away from electrodes (Weaker Electric Field)

Cell moves towards electrodes (Stronger Electric Field)

Fluid Flow   Electrode   Fluid Flow

FIGURE 3.3   Behaviour of nucleated biological cell in microchannel with excited electrodes

TABLE 3.1

Dielectric Properties of Various Biological Cells and Particles

Micro-entity	Diameter (μm)	Relative Permittivity	Electrical Conductivity (S/m)	Shell Size (nm)	Relative Permittivity of Shell	Electrical Conductivity of Shell (S/m)
Blood Cells						
RBCs [21]	5	59	0.31	9	4.44	1E-6
PLTs [22]	1.8	50	0.25	8	6	1E-6
Yeast Cells						
General Yeast Cells [23]	3.6	2657.7	0.25	8	2657.7	0.0016
Particles						
Polystyrene Particles [24]	1.8	2.5	2E-4	NA	NA	NA

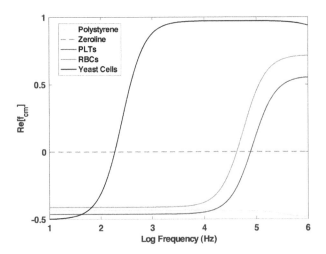

FIGURE 3.4 Results showing dielectric behaviour of various cells and polystyrene particles based on properties

for all the operating frequencies and yeast cells possesses only positive DEP values. To change the DEP effect in these two cases the only way is to change the medium conductivity and permittivity which may not be suitable as the biological suspension fluid come with limitations. In case of RBCs and PLTs both the DEP effects (positive and negative) come in different frequency ranges and so the forces on cells can be altered in the direction with just change in frequency of operation and not the medium.

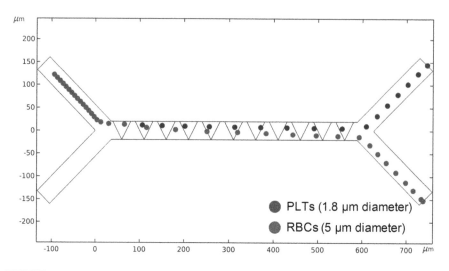

FIGURE 3.5   Behaviour of nucleated biological cell in microchannel with excited electrodes showing cell separation with only negative DEP effect and cell suspension is done from uniform initial position from the inlet on left side

3.4.1   Cell Response with Only Negative Dielectrophoresis

When cells/particles only possess the negative DEP effect, which means the $Re[f_{cm}]$ values remain negative throughout all the frequencies, then the initial cell position in the channel is important. In such case both type of cells should be suspended from the same initial opposition as shown in Figure 3.5. In this case cells get the same force polarity but different magnitude due to difference in size. The result shows that RBCs (in red) get more force and get collected in lower outlet, whereas PLTs received less force and got collected in upper outlet. The force acting in this case in negative which means direction of force is from up to down side. Opposite to this if the cells experience positive DEP, then the cells get attracted towards $Re[f_{cm}]$ the electrodes. We usually avoid the cases where only positive DEP effect is used for cell motion manipulation as there are chances that cell may get closer to electrodes where joule heating effects are high and cells may get destroyed.

3.4.2   Cell Response with Both Negative and Positive Dielectrophoresis

When the cells possess both DEP effect, which means the $Re[f_{cm}]$ values have negative as well as positive values, then selection of frequency and initial position of cells becomes important. Results for RBCs and yeast cells are shown in Figure 3.6 where cells are suspended from mid-point of the channel (from the left side of the channel in the figure) and get away from each other while moving in the channel from left to right. The cell in red are RBCs which are experiencing the negative DEP effect and cells shown in blue are yeast cells experiencing positive DEP effect at 100 kHz

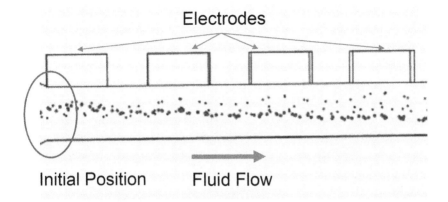

FIGURE 3.6 Simulation results showing behaviour of RBCs (in red) and yeast cells (in blue) in microchannel with excited electrodes (2 V, 1 kHz) showing cell bifurcation with RBCs having negative DEP effect and yeast cells

frequency in DI water. The judgement of initial position is done after the calculation of DEP effect from dielectric response graph.

3.4.3 MACHINE LEARNING-BASED CELL COUNTING METHOD

Recent development in Artificial Intelligence (AI) and Machine Learning (ML) technologies have shown significant advances in analysing objects based on size, colour scheme and type of objects. These techniques are mostly used where some decision-making needs to be accomplished based on some classification. These technologies can also be used in cell analysis for counting, detection and efficiency calculation. Some of the results for the analysis of size calculation are shown in Figure 3.7 using python programming. In this, the reference radius is taken and cells are allowed to be observed via images. The results have shown accurate detection of the targeted cells which can be further improved with an efficient mechanism in this domain.

3.5 CONCLUSIONS

Blood cell analysis for healthcare data extraction for diagnostic application is important for which the main process includes blood processing, cell separation, cell detection and cell counting. There are various methods available based on optical methods, magnetic methods, inertial forces and electrical methods. The dielectrophoresis is convenient and have no moving part associate, hence is preferred by many researchers. The cell separation is achieved with computational calculations, which includes the selection of medium parameters, dielectric response of the cells and frequency selection for effective separation. In this work, we have discussed the stratagem and computational method for cell separation using dielectrophoresis and

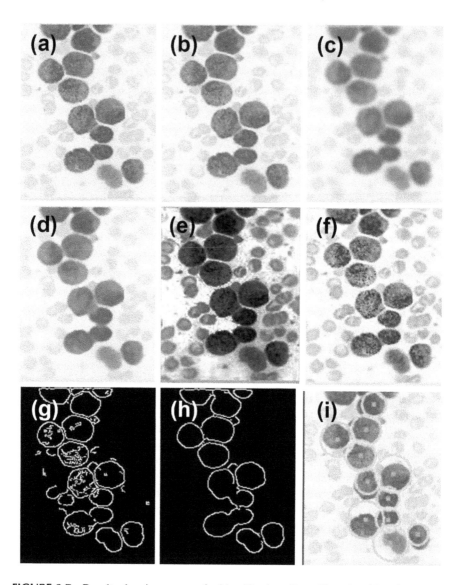

FIGURE 3.7 Results showing process for identification of specific using AI and ML technique in python, (a) original blood sample image [25], (b) processed grey scale image, (c) gaussian blur image, (d) median blur image, (e) histogram normalization, (f) CLAHE normalization, (g) any detailed edges detection, (h) detailed edges detection.

cell counting using AI and ML platform python. In cell separation method, the initial position of cells is important and cell experiencing similar force polarity should be suspended from one side of the channel only, and cells experiencing different polarity forces should be suspended from the middle of the channel that is central streamline. Apart from separation the cells can be counted using image processing based on edge identification post-grey scaling of the image.

REFERENCES

1. Müller, T., Gradl, G., Howitz, S., Shirley, S., Schnelle, T., & Fuhr, G. (1999). A 3-D microelectrode system for handling and caging single cells and particles. *Biosensors and Bioelectronics*, 14(3), 247–256.
2. Kumar, M., & Das, D. (2018, February). MEMS based flow cytometer with instrumentation system for detection of micro particles for health care applications. In *2018 Recent Advances on Engineering, Technology and Computational Sciences (RAETCS)* (pp. 1–5). IEEE.
3. Das, D., Biswas, K., & Das, S. (2014). A microfluidic device for continuous manipulation of biological cells using dielectrophoresis. *Medical Engineering and Physics*, 36(6), 726–731.
4. Gawad, S., Schild, L., & Renaud, P. H. (2001). Micromachined impedance spectroscopy flow cytometer for cell analysis and particle sizing. *Lab on a Chip*, 1(1), 76–82.
5. Zhang, H., Chon, C. H., Pan, X., & Li, D. (2009). Methods for counting particles in microfluidic applications. *Microfluidics and Nanofluidics*, 7(6), 739–749.
6. Holmes, D., Pettigrew, D., Reccius, C. H., Gwyer, J. D., van Berkel, C., Holloway, J., Davies, D. E., & Morgan, H. (2009). Leukocyte analysis and differentiation using high speed microfluidic single cell impedance cytometry. *Lab on a Chip*, 9(20), 2881–2889.
7. Johnson, P. T. (1983). Diseases caused by viruses, rickettsiae, bacteria, and fungi. *The Biology of the Crustacea Pathobiology*, 6, 2–78.
8. Hingorani, S. R., Petricoin III, E. F., Maitra, A., Rajapakse, V., King, C., Jacobetz, M. A., Ross, S., Conrads, T. P., Veenstra, T. D., Hitt, B. A., Kawaguchi, Y., Johann, D., Liotta, L. A., Crawford, H. C., Putt, M. E., Jacks, T., Wright, C. V., Hruban, R. H., Lowy, A. M., & Tuveson, D. A. (2003). Preinvasive and invasive ductal pancreatic cancer and its early detection in the mouse. *Cancer Cell*, 4(6), 437–450.
9. Fu, A. Y., Spence, C., Scherer, A., Arnold, F. H., & Quake, S. R. (1999). A microfabricated fluorescence-activated cell sorter. *Nature Biotechnology*, 17(11), 1109–1111.
10. Miltenyi, S., Müller, W., Weichel, W., & Radbruch, A. (1990). High gradient magnetic cell separation with MACS. *Cytometry: The Journal of the International Society for Analytical Cytology*, 11(2), 231–238.
11. Shi, Y., Ye, P., Yang, K., Meng, J., Guo, J., Pan, Z., Zhao, W., & Guo, J. (2021). Application of centrifugal microfluidics in immunoassay, biochemical analysis and molecular diagnosis. *Analyst*, 146(19), 5800–5821.
12. Xiang, N., & Ni, Z. (2022). Inertial microfluidics: Current status, challenges, and future opportunities. *Lab on a Chip*, 22(24), 4792–4804.
13. Choi, S., Song, S., Choi, C., & Park, J. K. (2007). Continuous blood cell separation by hydrophoretic filtration. *Lab on a Chip*, 7(11), 1532–1538.
14. Hawkes, J. J., & Coakley, W. T. (2001). Force field particle filter, combining ultrasound standing waves and laminar flow. *Sensors and Actuators Part B: Chemical*, 75(3), 213–222.
15. Kim, J., Erath, J., Rodriguez, A., & Yang, C. (2014). A high-efficiency microfluidic device for size-selective trapping and sorting. *Lab on a Chip*, 14(14), 2480–2490.
16. Kumar, M., Kumar, A., George, S. D., & Singh, K. (2021). A novel microfluidic device with tapered sidewall electrodes for efficient ternary blood cells (WBCs, RBCs and PLTs) separation. *Measurement Science and Technology*, 32(11), 115106.
17. Cho, S. H., Chen, C. H., Tsai, F. S., & Lo, Y. H. (2009, September). Micro-fabricated fluorescence-activated cell sorter. In *2009 Annual International Conference of the IEEE Engineering in Medicine and Biology Society* (pp. 1075–1078). IEEE.
18. Wang, Z., Wang, H., Lin, S., Ahmed, S., Angers, S., Sargent, E. H., & Kelley, S. O. (2022). Nanoparticle amplification labeling for high-performance magnetic cell sorting. *Nano Letters*, 22(12), 4774–4783.

19. Wang, Q., Zhang, X., Yin, D., Deng, J., Yang, J., & Hu, N. (2020). A continuous cell separation and collection approach on a microfilter and negative dielectrophoresis combined chip. *Micromachines*, 11(12), 1037.
20. Pohl, H. A., & Crane, J. S. (1971). Dielectrophoresis of cells. *Biophysical Journal*, 11(9), 711–727.
21. Piacentini, N., Mernier, G., Tornay, R., & Renaud, P. (2011). Separation of platelets from other blood cells in continuous-flow by dielectrophoresis field-flow-fractionation. *Biomicrofluidics*, 5(3), 034122.
22. Egger, M., & Donath, E. (1995). Electrorotation measurements of diamide-induced platelet activation changes. *Biophysical Journal*, 68(1), 364–372.
23. Kumar, M., Palekar, N., Kumar, A., & Singh, K. (2020). Optimized hydrodynamic focusing with multiple inlets in MEMS based microfluidic cell sorter for effective bio-cell separation. *Physica Scripta*, 95(11), 115005.
24. Choi, S., & Park, J. K. (2005). Microfluidic system for dielectrophoretic separation based on a trapezoidal electrode array. *Lab on a Chip*, 5(10), 1161–1167.
25. Bhamare, M. G., & Patil, D. S. (2013). Automatic blood cell analysis by using digital image processing: A preliminary study. *International Journal of Engineering Research and Technology (IJERT)*, 2(9), 3137–3141.

4 Efficient and Robust Digital Notch Filter Implementation Employing Wave Digital Structure

Abhay Sharma and Mridul Gupta

4.1 INTRODUCTION

In many signal-processing applications, retaining or deleting a certain frequency component from a signal while preserving the narrow or wide signal is a typical objective. This is a common use for peak and notch filters, which also have various uses in fields including telecommunication, automation, medical technology and speech processing [1–7]. It is possible to build any digital filters with either a finite or an infinite impulse response system. The finite impulse response systems are naturally stable and provide a phase linear output. In instances when phase linearity is not an absolute necessity, IIR implementation remains the technique of choice. This is because a filter with infinite impulse response can deliver an appropriate filter response with far less memory and fewer processing units as compared to a filter with finite impulse response can manage to achieve in order to do the same task [8–11].

When it comes to the effectiveness of the filtering process, the architecture of the digital filter plays a significant role, which results in hardware that is faster, smaller and less costly. The filters based on wave theory, which are closely connected to classical filter networks, are one of the most effective ways to create recursive digital filters [12]. Each wave digital filter (WDF) has a matching analogue domain reference filter from which it is generated [13, 14]. WDFs also solve IIR filter drawbacks such as sensitivity to word length as well as rounding mistakes in coefficients, all of which make IIR filter implementation a difficult challenge [15]. In systems needing digital filters, many WDFs are available and used. WDFs are known to have a number of beneficial characteristics, including a low coefficient susceptibility, a broad dynamic range and good stability. Wave digital filters may be further classified into a subcategory known as lattice wave digital filters (LWDF) [16, 17]. They are generated by converting analogue reference lattice filters to the digital domain and are renowned

DOI: 10.1201/9781003452645-4

for their very low passband sensitivity [18, 19]. These digital filters may be observed as a link between all-pass filters. Every all-pass branch is produced by a creative cascade of appropriately selected wave adaptors [17, 20]. The resultant structures are very modular, making them appropriate for signal processors and VLSI design.

This chapter aims to give a straightforward method for designing and implementing multiplier-less LWDF-based notch/peak filters. This work is separated into two halves. In the first section, all-pass filters are used to create the first, second and multiple order notch/peak filters. All-pass filters are replaced by adaptors, which are a chained realization of the second and first-order all-pass sections. When it comes to the number of multipliers that are necessary for a certain filter order, the use of adaptors results in an implementation that is both effective and efficient [21–26].

Complimentary filters, notch and peak can be obtained with a minimal amount of multipliers and an efficient digital wave structure. In the second section, the hardware implementation of these filters is performed on an FPGA Xilinx system generator for DSP EDA tool. The first- and second-order constant coefficient all-pass sections are produced using Richards' structures described in Ohlsson (2003) and Ohlsson and Wanhammar (2001) [27, 28]. These structures serve as the basis for notch filter realization. Adders, multipliers and delays are built into the first- and second-order adaptors.

4.2   NOTCH FILTER

The following response of frequency gives the notch filter ideal characteristics:

$$N_{ideal}\left(e^{j\omega}\right) = \begin{cases} 0, & \omega = \omega_0 \\ 1, & \text{otherwise} \end{cases}$$

where ω_0 is the notch frequency of the filter. The realization of notch filter using the function with all-pass behaviour is described as:

$$N(z) = \frac{1}{2}\left[1 + A(z)\right]$$

where $A(z)$ is the function with all-pass behaviour. At all frequencies, such a function generates a unity frequency magnitude response.

The equation that describes the transfer function of a notch filter that uses first-order all-pass is as follows:

$$N(z) = \frac{1}{2}\left[1 + \frac{A_1(z)}{z}\right]$$

Where,

$$A_1(z) = \frac{k_1 + \dfrac{1}{z}}{1 + k_1\dfrac{1}{z}}$$

here $k_1 = -\cos\omega_0$ and multiplication by z^{-1} modifies the second-term phase from 0 to -2π. A first-order notch filter's rejection bandwidth is set at $\pi/2$ and it does not depend on the notch frequency [7]. The notch filter may be accomplished using a second-order all-pass filter that provides frequency and bandwidth control for the notch.

It can be shown that the transfer function of a notch filter that uses an all-pass function of second order is represented by:

$$N(z) = \frac{1}{2}\left[1 + A_2(z)\right]$$

where

$$A_2(z) = \frac{K_2 + K_1\left(1 + K_2\right)\dfrac{1}{z} + \dfrac{1}{z^2}}{1 + K_1\left(1 + K_2\right)\dfrac{1}{z} + K_2\dfrac{1}{z^2}}$$

Using the following relations, the second-order all-pass function-based notch filter permits independent setting of the cut-off frequency (ω_0) and notch-width (Ω).

$$K_1 = -\cos\omega_0$$

$$K_2 = \frac{1 - \tan\dfrac{\Omega}{2}}{1 + \tan\dfrac{\Omega}{2}}$$

The basic implementation structure of all-pass function-based notch filter is shown in Figure 4.1.

Filters with multiple notches are effective for decreasing harmonics of power line interference from ECG signal recordings. Harmonics are caused by the non-linear characteristics of the transformer cores found in the power supply. Several second-order notch filters may be cascaded to create a high-order multiple notch filter [5]. Filter with multiple notches can be described as:

$$N(z) = \prod_{j=1}^{L} \frac{1}{2}\left(1 + A_{2j}\right)$$

On expansion:

$$N(z) = \prod_{j=1}^{L} \frac{1}{2}\left(1 + \frac{K_{2j} + K_{1j}\left(1 + K_{2j}\right)\dfrac{1}{z} + \dfrac{1}{z^2}}{1 + K_{1j}\left(1 + K_{2j}\right)\dfrac{1}{z} + K_{2j}\dfrac{1}{z}}\right)$$

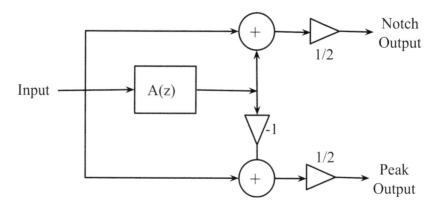

FIGURE 4.1 Filter implementation using all-pass function

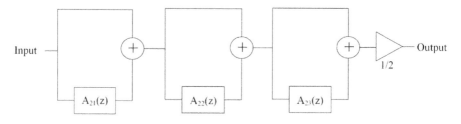

FIGURE 4.2 Triple notch filter implementation

here, j represents the number of different notches to be generated. As an example, triple-notch filter is achieved by cascading three second-order notch filters as depicted in Figure 4.2.

4.3 WAVE DIGITAL STRUCTURE

As opposed to voltage and current as signal variables, the travelling wave quantities are used for establishing the relationship between a filter in the wave domain and its reference network [14]. It is possible to define each n-port analogue network by making use of the concepts of incident and reflected wave values, which are connected to current I, voltage V and port resistance R as respective variables.

$$A = V + IR$$

$$B = V - IR$$

where quantities representing the incident and reflected wave are A and B, respectively. R denotes the resistance of the port, V indicates the port voltage and I refers to the port current. [14]. The above equations can be used to show any analogue element as a two-port network. Adaptors are used to connect two

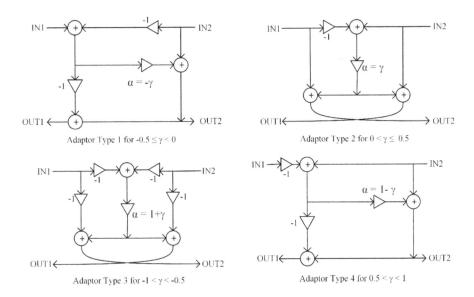

FIGURE 4.3   Adapter types and their selection based on value of γ

ports of a network to each other. A two-port adapter is just a digital all-pass section with two ports. Lattice wave digital filters are made up of a series of second- and first-order all-pass functions. The coefficient value of each adapter determines the response of the all-pass section, which regulates the response. For an adaptor, computational comprises one multiplier and three additions. As lattice wave building blocks, there are four types of adaptors [17, 18]. Figure 4.3 shows the four single multiplier two-port configuration. All four adaptors are the same because they all use the same relation:

$$b_1 = \gamma a_1 + (1+\gamma)a_2$$

$$b_2 = (1-\gamma)a_1 - \gamma a_2$$

where γ is the wave coefficient, whose value lies in the range −1 to +1. The architecture of each adapter is decided depending on the value of γ as depicted in Figure 4.3.

The coefficient value of the genuine multiplier (α) will always be +ve and either less than or equal to half $\left(0 \le \alpha \le \dfrac{1}{2}\right)$ in these structures.

First-order two-port adapter and delay element form Richards' all-pass section. Figure 4.4 shows this conventional structure's signal flow graph. As depicted in Figure 4.5, the Richards' all-pass section of order 2 is generated by connecting the two sections of order 1.

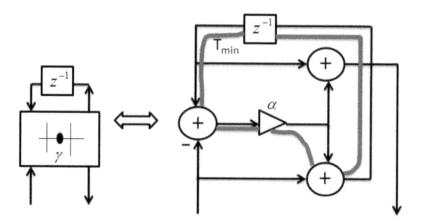

FIGURE 4.4   Richards' all-pass section of order 1

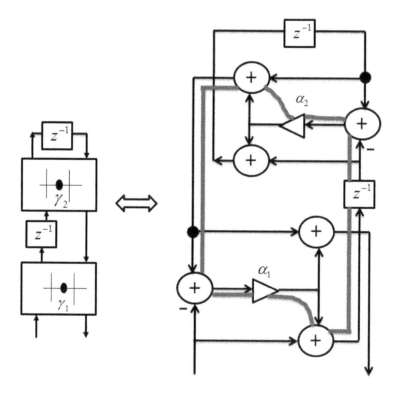

FIGURE 4.5   Richards' all-pass section of order 2

4.4   WAVE DIGITAL NOTCH FILTER

For the purpose of illustrating the process of designing notch filters utilizing wave digital structure, two design examples are given:

Example 1: First-order notch filter with notch frequency $\omega_0 = 0.49\pi$. By using the successive Schur parametrization approach, one may determine the transfer function for a wave digital notch filter. This can be done by calculating the coefficients of the multiplier [17] (Figure 4.6).

$$N(z) = \frac{1}{2}\left[1 + \frac{1}{z}\frac{-0.03131 + \frac{1}{z}}{1 - 0.03131\frac{1}{z}}\right]$$

Example 2: Digital specification of notch filter with order 2 are: $\omega_0 = 0.3776\pi$ and $\Omega = 0.06559\pi$. It is possible to provide an expression for the all-pass function of order 2 in terms of the adaptor coefficient as: (Figure 4.7)

$$A(z) = \frac{-\gamma_{2j-1}z^2 + \gamma_{2j}\left(\gamma_{2j-1} - 1\right)z + 1}{z^2 + \gamma_{2j}\left(\gamma_{2j-1} - 1\right)z - \gamma_{2j-1}}$$

4.5   FPGA IMPLEMENTATION

Several platforms existed which are utilized for realization in hardware of signal-processing architectures. Field programmable gate arrays (FPGA) are the

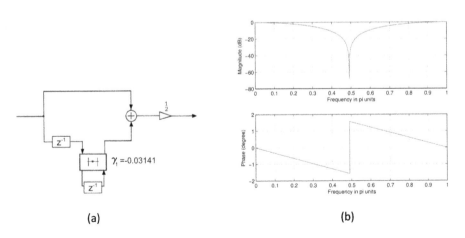

(a) (b)

FIGURE 4.6   Design Example 1: (a) Block diagram, (b) Response (magnitude and phase)

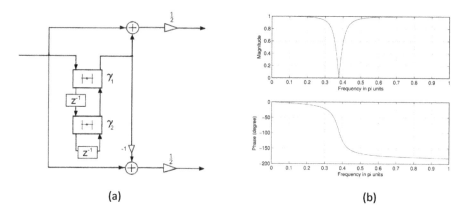

(a) (b)

FIGURE 4.7 Design Example 2: (a) Block diagram, (b) Response (magnitude and phase)

preferred option. This is the case because the system has programmable logic blocks as well as connections that may be rearranged. In this instance, the electronic device automation tool System Generator for DSP by Xilinx is used in order to carry out the hardware realization of the recommended notch filter through wave digital structure. This cutting-edge piece of software makes it possible to include the Xilinx block set into the MATLAB Simulink library. The Xilinx blocks are pre-existing Ips. Their specifications can be modified according to the requirement of the design. Therefore, systems in the digital domain may be built on FPGA using models based on Simulink. All system generator models need the presence of a Gateway In, Gateway Out and a System Generator token. Data conversion from the Simulink environment to the fixed or floating-point representation of the FPGA is handled by the Gateway In and Gateway Out blocks. All of the digital system's input and output ports are determined by Gateway blocks. (after synthesis or after implementation). Due to the Xilinx watermark, System Generator blocks stand apart from the rest of Simulink's standard library (Figure 4.8).

The design stages for the EDA tool Xilinx System Generator for DSP are as follows:

- Design submission: a Simulink model that includes Xilinx blocks in addition to Simulink blocks (optional).
- Verification of design: Simulation in Simulink environment will be used for verifying the design.
- Compilation of design: Conversion of model to RTL netlist employing pre-existing IP configured in user-defined parameters. At this point, the resource or the timing analyser may be triggered.
- The production of FPGA bitstream files for programming, as well as the instantiation of hardware co-simulation, if necessary.

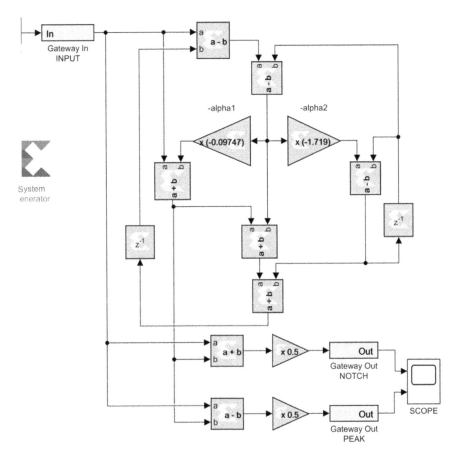

FIGURE 4.8   FPGA implementation using Xilinx System Generator for DSP

4.6   CONCLUSION

Notch filters are widely used in digital systems and hence in this chapter an efficient hardware realization using wave digital architecture is presented. The primary
criteria for digital filter implementation are that they have a low complexity and a
fast speed. In the presented wave digital notch filter design, the idea of an all-pass
filter design is applied. The filters that have been suggested are regular and modular,
and they have the capacity to concurrently realize a number of different transfer
functions. Further, it is illustrated that Xilinx system generator can be exploited for
FPGA implementation in a straightforward block-level design.

REFERENCE

1. J. C. Huhta and J. G. Webster, 60-Hz interference in electrocadiographn, *IEEE
   Transactions on Bio-Medical Engineering*, 20(2) (1973) 91–101.

2. C. D. McManus, D. Neubert and E. Cramer, Characterization and elimination of AC noise in electrocardiograms: A comparison of digital filtering methods, *Computers and Biomedical Research, An International Journal*, 26(1) (1993) 48–67.
3. K. Hirano, S. Nishimura and S. K. Mitra, Design of digital notch filters, *IEEE Transactions on Circuits and Systems, CAS-21* (Jul. 1974) 540–546.
4. Jonathan S. Abel and David P. Berners, Filter design using second-order peaking and shelving sections, *Proceedings of ICMC*, 26(1) (2004) University of Michigan Library.
5. Soo-Chang Pei and Tseng Chien-Cheng, IIR multiple notch filter design based on Allpass filter, *IEEE Transactions on Circuits and Systems. Part II: Analog and Digital Signal Processing*, 44(2) (Feb. 1997) 133–136.
6. S. C. Dutta Roy, Balbir Kumar and Shail Bala Jain, FIR notch filter design-A review, electronics and energetics, 14(3) (2001) 295–327.
7. Y. V. Joshi and S. C. Dutta Roy, Design of IIR digital notch filters, *Circuits, Systems, and Signal Processing*, 16(4) (1997) 415–427.
8. S. Harize, M. Benouaret and N. Doghmane, A methodology for implementing decimator FIR filters on FPGA, *AEU – International Journal of Electronics and Communications*, 67(12) (Dec. 2013) 9931004.
9. R. I. Hartley, Subexpression sharing in filters using canonic signed digit multipliers, *IEEE Transactions on Circuits and Systems. Part II*, 43(10) (Oct. 1996) 677–688.
10. R. Mahesh and A. P. Vinod, A new common subexpression elimination algorithm for realizing low complexity higher order digital filters, *IEEE Transactions on Computer-Aided Design of Integrated Circuits and Systems*, 27(2) (Feb. 2008) 217–219.
11. Ljiljana D. Milic and Miroslav D. Lutovac, Efficient algorithm for the design of high-speed elliptic IIR filters, *AEU - International Journal of Electronics and Communications*, 57(4) (2003) 255–262.
12. V. M. Poucki, A. Zemva, M. D. Lutovac and T. Karcnik, Elliptic IIR filter sharpening implemented on FPGA, *Digital Signal Processing*, 20(1) (2010) 13–22.
13. A. Fettweis, Digital filter structures related to classical filter networks, *Arch. Elektron. and Uebertragungstech*, 25 (1971) 19–89.
14. A. Fettweis, Wave digital filters: Theory and practice, *Proceedings of the IEEE*, 74(2) (1986) 270–327.
15. Andreas Antoniou, *Digital Signal Processing*, McGraw-Hill Publication, 2006.
16. A. Fettweis, H. Levin and A. Sedlrneyer, Wave digital lattice filters, *International Journal of Circuit Theory and Applications*, 2(2) (Jun. 1974) 203–211.
17. L. Gazsi, Explicit formulas for lattice wave digital filters, *IEEE Transactions on Circuits and Systems*, 32(1) (1985) 68–88.
18. J. Yli-Kaakinen and T. Saramaki, A systematic algorithm for the design of lattice wave digital filters with short-coefficient wordlength, *IEEE Transactions on Circuits and Systems. Part I*, 54(8) (Aug. 2007) 1838–1851.
19. Timo I. Laakso, Jari Ranta and Seppo J. Ovaska, Design and implementation of efficient IIR notch filters with quantization error feedback, *IEEE Transactions on Instrumentation and Measurement*, 43(3) (Jun. 1994) 449–456.
20. Chien-Cheng Tseng and Soo-Chang Pei, Stable IIR notch filter design with optimal pole placement, *IEEE Transactions on Signal Processing*, 49(11) (Nov. 2001) 2673–2681.
21. P. A. Regalia, S. K.Mitra and P. P.Vaidyanathan, The digital all-pass filter: A versatile signal processing building block, *Proceedings of the IEEE*, 76(1) (Jan. 1988) 19–37.
22. J. F. Chicharo and T. S. Ng, Gradient-based adaptive IIR notch filtering for frequency estimation, *IEEE Transactions on Acoustics, Speech, and Signal Processing*, 38(5) (May 1990) 769–777.
23. S. K. Mitra, *Digital Signal Processing*, 4th edition, McGraw Hill, 2013.

24. P. P. Vaidyanathan, S. K. Mitra and Y. Neuvo, A new approach to the realization of low sensitivity IIR digital filters, *IEEE Transactions on Acoustics, Speech, and Signal Processing, ASSP-34,* 2 (Apr. 1986) 350–361.
25. M.Yaseen, *Direct Design of Bandtop Wave Digital Lattice Filter, ICECS,* 2010, 847–850.
26. S. A. Samad, A. Hussain and D. Isa, Wave digital filters with minimum multiplier for discrete Hilbert transformer realization, *International Journal of Signal Processing,* 86(12) (2006) 3761–3768.
27. H. Ohlsson, Studies on implementation of digital filters with high throughput and low power consumption, Thesis No. 1031, Linkping studies in science and technology, 2003.
28. H. Ohlsson and L. Wanhammar, Implementation of bit-parallel lattice wave digital filters, *Proceedings of the Swedish System-on-Chip Conference* (Mar. 2001) 71–74.

5 Computationally Efficient Fractional-order Filters

Shalabh K. Mishra, Om P. Goswami,
Nawnit Kumar and Naveen Kumar

5.1 INTRODUCTION

Filters are one of the most popular and widely used electronic devices that are used in signal selection, signal rejection and phase transformation [1, 2]. Nowadays, digital systems are gaining more attention over the continuous time systems due to some inherent merits such as accuracy, design flexibility, noise immunity, etc. Digital filters consist of multipliers, adders and delay circuits in lieu of resistive, capacitive and/or inductive circuit components [3–9]. Electronic filters are used in several areas of electrical engineering and biomedical sciences such as communication systems, signal processing, image and speech processing, radar, GPS navigation systems, etc.

The performance of any electronic filter is characterized by its pass band, roll-off rate and the stop band. An ideal filter should have a very high roll-off rate and therefore sharp transition occurs between pass band and stop band. Although ideal filter does not exist practically, they may be approximated by increasing the order of the system. Furthermore, the higher-order filters have their own limitations. Higher-order filters are more complex, less stable and may have higher settling time. Therefore, the selection of order is a very crucial step and it should be chosen very wisely to maintain a justified trade-off between selectivity (roll-off rate) and settling time and stability. Thus the selection of order is a brainstorming task for the designing of efficient filters. To resolve this issue, researchers are now applying the concept of fractional calculus in filter designing; which gives us more freedom in the selection of order and tune the system performance very neatly; such filters are defined as fractional-order filters [10–12]. Generalized non-integer-order filters have added constructional liberty and flexibility as compared to traditional integer-order filters. Besides this, fractional calculus is also applied in several other areas of science and engineering such as analogue circuits, control systems, power electronics, electromagnetics, biomedical engineering, etc.

Digital filters are designed using the two popular topologies: FIR and IIR designing. Each of these techniques has their own separate merits and limitation.

The IIR filter is often preferred over the FIR since the circuit is less complex for similar specifications. For the IIR designing scheme, filters are initially designed in

DOI: 10.1201/9781003452645-5

continuous time domain and then appropriate s-to-z transformations are applied to transfer this analogue function in the digital domain.

Numerous such transformations are available that can be applied to replicate an analogue filter in the digital domain. Some popular mapping functions are bilinear transformation, fractional-bilinear transformation, Al-Alaoui operator, Schneider operator, Admas- third-order, etc. [10, 13–20]. The bilinear is long-established s-to-z transforms since it gives the first-order, stable mapping, although it has one major drawback which is known as frequency *warping*. Due to the warping effect the designed IIR filter is unable to imitate the analogue system accurately in the discrete time domain. Although this warping effect may be diminished using the frequency pre-warping, several drawbacks have also been reported of this pre-warping concept [10, 21]. For example, the pre-warping scheme preserves the pass-band and stop-band frequency only, but is unable to retain the actual shape of the magnitude response curve. Moreover, this scheme is applicable to the selected type of specific filters only, and therefore can't be applied in fractional order filters. Recently Mishra et al. developed a first-order mapping function using the *Particle Swarm Optimization* (PSO) algorithm [21]. Besides this, there is one more approach to implementing the IIR filter, in which a generalized transfer function is considered and their coefficients are obtained by using the metaheuristic stochastic optimization techniques. The next section discusses the detailed procedure to formulate and realize an IIR filter with suitable examples.

5.2 IIR FILTER DESIGNING

In the conventional IIR designing scheme, a continuous time filter is directly transferred into the digital domain using the s-to-z transformation. Some popular s-to-z mapping transformations are tabulated in Table 5.1. The magnitude and phase response of the s-to-z mapping functions are depicted in Figure 5.1. It is observed that the bilinear and optimized bilinear are capable of maintaining a phase angle of $\pi/2$ but fail to provide satisfactory magnitude mapping within the complete frequency; whereas Al-Alaoui and other elevated order mapping functions provide sufficient resemblance in terms of magnitude only, but are unable to hold desirable phase mapping. It is well proven that the phase as well as magnitude resemblances is equally important for an optimal s-to-z mapping function used in IIR designing [10, 21]. Since the second-order Schneider operator possesses a justified balance between magnitude and phase mapping, it may be preferred for IIR designing.

5.2.1 DESIGN EXAMPLE

Consider a fractional-order low-pass filter expressed by equation (5.1), in which $g_1 = 1$, $g_2 = 1.31$, $g_3 = 0.99$ and $\alpha = 0.9$.

$$H(s) = \frac{g_1}{s^{1+\alpha} + g_2 s^{\alpha} + g_3} \tag{5.1}$$

TABLE 5.1

Mapping Function for Continuous time to Discrete time Mapping

S. No.	Transformation	Mapping Function
1	Bilinear	$s = \dfrac{2(z-1)}{T(z+1)}$
2	Al-Alaoui	$s = \dfrac{8(z-1)}{7T(z+1/7)}$
3	Optimized Bilinear	$s = \dfrac{(z-1)}{0.527T(z+0.997)}$
4	Schneider	$s = \dfrac{12(z^2 - z)}{T(5z^2 + 8z - 1)}$
5	Admas third-order	$s = \dfrac{24(z^3 - z^2)}{T(9z^3 + 19z^2 - 58z + 1)}$
6.	Jalloul-Aloui	$s = \dfrac{\left(z^3 - 0.0259z^2 - 0.7763z - 0.1964\right)}{T\left(0.8656z^3 + 0.9969z^2 + 0.2930z + 0.0162\right)}$

This filter is designed in the digital domain using an IIR realization scheme using some popularly used s-to-z transformations. The magnitude response of the considered IIR filter is shown in Figure 5.2. Although the Al-Alaoui and the third-order Jalloul-Aloui operators give satisfactory magnitude mapping of s-to-z mapping, but fail to provide satisfactory magnitude response of the digital counterpart of the analogue Butterworth filter. The optimized bilinear gives better results as compared to

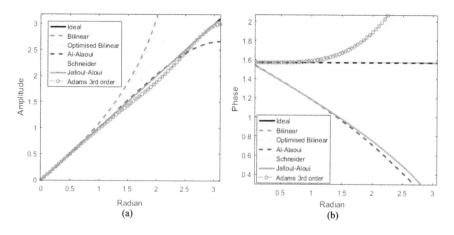

FIGURE 5.1   Magnitude and phase response of the s-to-z mapping functions

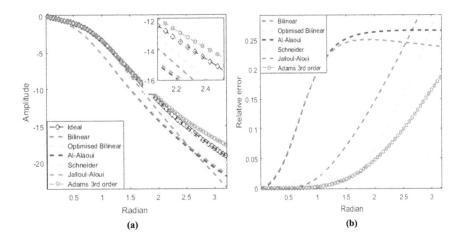

FIGURE 5.2   Magnitude response and relative error

normal bilinear transformation whereas the second-order Schneider gives the most accurate result among the considered mapping functions.

5.2.2   STABILITY ANALYSIS

First-order s-to-z mapping functions provide one-to-one mapping between the s-plane and z-plane. Therefore, mapping of the s-plane with the z-plane is unique, and it is possible to locate the poles of the designed IIR filter if the coordinates of analogue poles (s-plane) are already known. Furthermore, the bilinear, optimized bilinear and Al-Alaoui, operator maps the left half of the s-plane with the area inside the unit circle of the z-plane, which means the stable region of the analogue domain is always mapped with the stable region of the digital domain only. So, a stable analogue filter (s-domain) will always give stable IIR transformation if the design uses these mapping functions.

Contrary to this, the higher-order s-to-z mapping functions give one-to-many mapping of the s-plane with the z-plane, which means mapping of the s-plane to the z-plane is not unique. This type of mapping may map the stable region of the s-plane with the unstable region of the z-plane and therefore, leads to serious stability issues. Hence, the IIR filter designed using the higher-order mapping functions must be passed through the stability check before the final realization of the system. This stability check can be done through the pole-zero mapping in the z-plane, and the system will be stable if no poles lie outside the unit circle.

5.3   DIRECT IMPLEMENTATION USING OPTIMIZATION

In the past decade, various approaches based on metaheuristic optimizations have been employed to approximate the filter design in the digital domain [22, 23]. The generalized transfer function of an IIR filter is:

$$H(z) = \frac{\sum\limits_{k=0}^{M} b_k z^{-k}}{\sum\limits_{k=0}^{M} a_k z^{-k}} \tag{5.2}$$

For the second-order system, the generalized transfer function is:

$$H_1(z) = \frac{a_0 z^2 + a_1 z + a_2}{b_0 z^2 + b_1 z + b_2} \tag{5.3}$$

For the third-order system, the generalized transfer function is:

$$H_2(z) = \frac{a_0 z^3 + a_1 z^2 + a_2 z + a_3}{b_0 z^3 + b_1 z^2 + b_2 z + b_3} \tag{5.4}$$

The coefficient of these transfer functions can directly be optimized to get the magnitude approximation of the low-pass filter in the generalized fractional sense. The L_1-norm fitness function comes up with a constant magnitude response, less overshoot and few ripples at the disrupted points [24]. Therefore, this error objective function is applied to reduce the dissimilarity between the ideal (Analog) and the approximated magnitude response. For second and third order the error functions E_1 and E_2, respectively, are as given below.

$$E_1 = \left\| H(s) \right| - \left| H_1(z) \right|_{z=e^{j\omega}} \right| \tag{5.5}$$

$$E_2 = \left\| H(s) \right| - \left| H_2(z) \right|_{z=e^{j\omega}} \right| \tag{5.6}$$

The different modern nature-inspired optimization techniques utilized for optimizing L1-norm-based error objective functions are described in this section. The error functions E_1 and E_2 are minimized using Grey Wolf Optimize (GWO), Ant Lion Optimizer (ALO) and Multiverse Optimizer (MVO) techniques and evaluated for 500 iterations to get the best possible outcome. Eventually, the selection criteria of the controlling parameters decide the excellence of any nature-inspired algorithm for a specific problem. Additionally, the internal parameters including the number of iterations and the count of population may be regulated to achieve the optimal results as they provide a justified trade-off between the global search phase and local search. Some popular optimization techniques used for designing of FOLPFs are briefly discussed below.

5.3.1   THE ANT LION OPTIMIZATION

The ALO algorithm is a metaheuristic optimization algorithm developed from the hunting nature of antlions. Antlions are insects that dig conical pits in the sand to trap ants, which they then prey upon. The ALO algorithm consists of two main components: the ant component and the lion component. The ant component represents the exploration phase, in which a set of virtual ants move around the search space to find potential solutions. The lion component represents the exploitation phase, in which a virtual lion waits at the bottom of the conical pit to capture the ant that falls in. During the exploration phase, each ant moves around the search space using a random walk strategy, guided by a pheromone trail. The pheromone trail is updated based on the fitness of the solutions found by the ants. This encourages the ants to explore regions of the search space that have produced good solutions in the past. During the exploitation phase, the virtual lion captures the best ant found during the exploration phase and makes small adjustments to its solution to improve its fitness. This process is repeated for a set number of iterations or until a termination criterion is met. The ALO algorithm has been shown to be effective in resolving diverse problems based on optimization, such as design problems, scheduling problems and image-processing problems. It is particularly well-suited for problems that have multiple local optima and complex search spaces [25, 26].

5.3.2　Grey Wolf Optimizer

The GWO is a metaheuristic optimization algorithm inspired by the social hierarchy and hunting behaviour of grey wolves [27, 28]. The GWO algorithm simulates the hunting nature of grey wolves, which includes four steps: searching the prey, encircling the prey, attacking the prey and feeding on it. In the context of optimization, the prey corresponds to the optimal solution of a problem, and the wolves correspond to the search agents. At each iteration of the algorithm, the positions of the search agents (wolves) are updated based on their current positions and the positions of the δ, θ and Φ wolves, which represent the best, second best and third best results found so far. The positions are updated using a set of equations that simulate the movement of the wolves towards the prey. The GWO algorithm appears to be efficacious in several versatile classes of optimization problems, including function optimization, engineering design and image processing. However, like all metaheuristic algorithms, its performance can depend on the specific problem being solved and the algorithm parameters.

5.3.3　Multiverse Optimizer Algorithm

The MVO algorithm is a metaheuristic optimization technique inspired by the concept of the multiverse in physics [29, 30]. The multiverse hypothesis suggests that there may be multiple universes or parallel realities, each with its own set of physical laws and constants. The MVO algorithm simulates the concept of the multiverse by creating multiple search spaces, each representing a different universe or reality. Each search space has its own set of solutions, and the algorithm attempts to find the optimal solution by exchanging information between the search spaces.

At the beginning of the optimization process, the algorithm creates a set of random solutions in each search space. The solutions are then evaluated based on their fitness, and optimal solutions in each search space are identified.

The algorithm then performs a series of operations to exchange information between the search spaces. These operations include the creation of new solutions in each search space based on information from other search spaces and the selection of the best solutions from each search space to be used in the next iteration. The MVO algorithm also incorporates a diversity maintenance mechanism, which encourages the exploration of different regions of the search space by penalizing solutions that are too similar to existing solutions. The MVO algorithm has been shown to be effective in solving a wide range of optimization problems, such as design problems, scheduling problems and data-mining problems. It is particularly well-suited for problems that have multiple local optima and complex search spaces, as it can effectively explore multiple regions of the search space simultaneously.

5.4   RESULTS AND DISCUSSION

The optimization and resultant transfer functions have been simulated using the MATLAB platform. The computed second- and third-order transfer function coefficients are displayed in Table 5.2. For the second- and third-order transfer functions, the Sum of Absolute Magnitude Error (SAME) has also been determined for each optimization technique. The approximated second- and third-order magnitude response of the analogue filter has been plotted in Figures 5.3 and 5.4, respectively. They approximate the ideal analogue magnitude response considerably well over the complete Nyquist interval. Figures 5.5 and 5.6 compare the absolute magnitude error of the different optimization techniques for the second- and third-order designs, respectively. As observed in Figure 5.5, third-order GWO and MVO-based designs offer reduced inaccuracy in the

TABLE 5.2

Transfer Functions and Statistical Comparison of Second and Third-order Designs

Order	Optimization Method	Numerator	Denominator	SAME
Second-order	GWO	[−0.47148, −1.5752, −0.44108]	[−0.73621, 3.8642, −0.63163]	3.0967
	ALO	[−1.3212, −0.7108, −0.11183]	[−3.4369, 1.5776, −0.32629]	1.9400
	MVO	[1.5004, 1.0303, 0.23213]	[−0.4194, 3.9941, −0.79709]	2.9188
Third-order	GWO	[−0.078556, −0.0069386, 1.0347, 1.5316]	[3.9977, −1.3039, −0.47153, 0.36541]	1.8765
	ALO	[−0.87745, −1.5553, −0.60716, −0.10959]	[−0.33014, 1.2125, −2.0542, −2.0516]	1.1242
	MVO	[−0.2309, −1.9921, −2.6836, −0.45956]	[3.7784, 3.7774, −2.9359, 0.92513]	0.8369

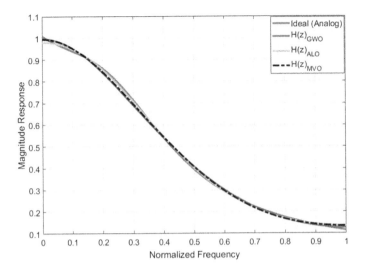

FIGURE 5.3   Response of second-order optimized system

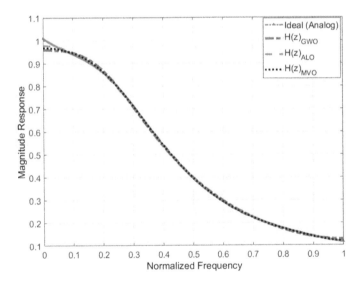

FIGURE 5.4   Response of third-order optimized system

low-frequency region while ALO produces better results for the whole Nyquist interval. The statistical comparison in Table 5.2 also demonstrates the fact that ALO offers the least SAME for the entire Nyquist range. In Figure 5.6, the third-order designs have further reduced the magnitude errors, and yield an error less than 0.015 for $0.1\pi \leq \omega \leq \pi$ and 0.01 for $0.2\pi \leq \omega \leq 0.95\pi$. Moreover, the design

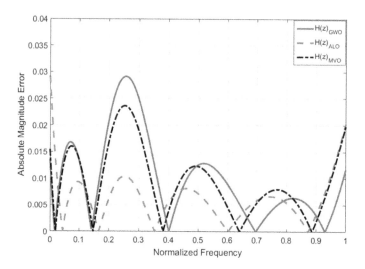

FIGURE 5.5 Magnitude error (Absolute) of second-order optimized system

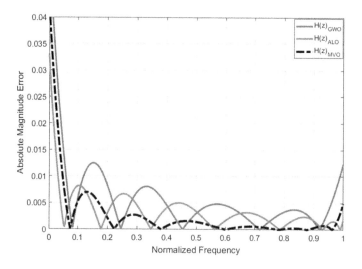

FIGURE 5.6 Magnitude error (Absolute) of third-order optimized system

obtained by MVO provides the least error in terms of the least magnitude error for the complete Nyquist range. The statistical comparison in Table 5.2 also supports the fact that MVO-based third-order design provides the least error and outperforms the other optimization algorithms comparatively.

Therefore, instead of employing the traditional s-to-z mapping functions, direct approximation of fractional analogue filters in the digital domain using optimization techniques may also be a good alternative for designing digital filters.

REFERENCES

1. R. Schaumann, H. Xiao, M. V. Valkenburg, *Design of Analog Filters*. Oxford University Press, 2001.
2. S. Venkateswaran, R. Singh, "A symmetric notch filter with two grounded capacitors," *IETE Journal of Research*, 20(1–2), pp. 1–6, 1974.
3. S. C. Dutta Roy, B. Tholeti, "A new feedback configuration for canonical realization of IIR filters and its application to lattice realizations," *IETE Journal of Research*, 54(1), pp. 45–50, 2008.
4. A. Thenmozhi, S. D. R. Prasath, S. Raju, V. Abhaikumar, "A novel approach to synthesize microwave filters using artificial neural networks," *IETE Journal of Research*, 54(2), pp. 105–109, 2014.
5. A. Antoniou, *Digital Filters: Analysis, Design, and Application*, 2nd edition. McGraw Hill Education, 2001.
6. S. K. Mitra, *Digital Signal Processing: A Computer-Based Approach*. New Delhi: Tata Mcgraw-hill Publishing Company Limited, 2001.
7. A. Aggarwal, T. K. Rawat, M. Kumar, D. K. Upadhyay, "Design of optimal band-stop FIR filter usingL1-norm based RCGA," *Ain Shams Engineering Journal*, 9(2), 2016. DOI: 10.1016/j.asej.2015.11.022
8. A. Aggarwal, T. K. Rawat, M. Kumar, D. K. Upadhyay, "Optimal design of FIR high pass filter based on L_1 error approximation using real coded genetic algorithm," *Engineering Science and Technology, An International Journal*, 18(4), pp. 594–602, 2015.
9. A. Aggarwal, T. K. Rawat, D. K. Upadhyay, "Design of optimal digital FIR filters using evolutionary and swarm optimization techniques," AEU - International Journal of Electronics and Communications, 70(4), pp. 373–385, 2016.
10. S. K. Mishra, D. K. Upadhyay, M. Gupta, "Search of optimal s-to-z mapping function for IIR filter designing without frequency prewarping," *IETE Journal of Research*, 2019. DOI: 10.1080/03772063.2019.1569484
11. T. Hélie, "Simulation of fractional-order low-pass filters," *IEEE/ACM Transactions on Audio, Speech, and Language Processing*, 22(12), pp. 1636–1647, 2014.
12. Z. Gao, "Fractional-order Kalman filters for continuous-time linear and nonlinear fractional-order systems using Tustin generating function," *International Journal of Control*, 2017. DOI: 10.1080/00207179.2017.1378438
13. M. A. Al-Alaoui, "Al-Alaoui operator and the new transformation polynomials for discretization of analogue systems," *Electrical Engineering*, 90(6), pp. 455–467, 2008.
14. F. Leulmi, Y. Ferdi, "Improved digital rational approximation of the operator $S\alpha$ using second-order s-to-z transform and signal modeling," *Circuits, Systems, and Signal Processing*, 34(6), pp. 1869–1891, 2014.
15. M. A. Al-Alaoui, "Filling the gap between the bilinear and the backward-difference transforms: An interactive design approach," *International Journal of Electrical Engineering Education*, 34(4), pp. 331–337, 1997.
16. S. C. Pei, H. J. Hsu, "Fractional bilinear transform for analog-to-digital conversion," *IEEE Transactions on Signal Processing*, 56(5), pp. 2122–2127, 2008.
17. A. Schneider, J. Kanshige, D. Groutage, "Higher order s-to-z mapping functions and their application in digitizing continuous-time filters," *Proceedings of the IEEE*, 79(11), pp. 1661–1674, 1991.
18. A. M. Schnieder, J. A. Anuskiewicz, I. S. Barghouti, "Accuracy and stability of discrete-time filters generated by higher-order s-to-z mapping functions," *IEEE Transactions on Automatic Control*, 39(2), pp. 435–441, 1994.
19. C. Wan, A. M. Schnieder, "Further improvements in digitizing continuous-time filters," *IEEE Transactions on Signal Processing*, 45(3), pp. 533–542, 1997.

20. M. A. Al-Alaoui, "Novel stable higher order s-to-z transforms," *IEEE Transactions on Circuits and Systems Part I: Fundamental Theory and Applications*, 48(11), pp. 1326–1329, 2001.

21. S. K. Mishra, D. K. Upadhyay, M. Gupta, "Optimized First-order s-to-z Mapping Function for IIR Filter Designing," *2019 20th International Conference on Intelligent System Application to Power Systems (ISAP)*, 2019. DOI: 10.1109/ISAP48318.2019.9065934

22. O. P. Goswami, T. K. Rawat, D. K. Upadhyay, "L1-norm-based optimal design of digital differentiator using multiverse optimization," *Circuits, Systems, and Signal Processing*, 41(8), pp 4707–4715, 2022.

23. O. P. Goswami, D. K. Upadhyay, T. K. Rawat, "Extended bilinear transform and multi-rate technique-based approach for analog-to-digital transform," *International Journal of Electronics*, 2021. https://doi.org/10.1080/00207217.2021.1969446

24. A. Aggarwal, T. K. Rawat, D. K. Upadhyay, "Optimal design of L1-norm based IIR digital differentiators and integrators using the bat algorithm," *IET Signal Processing*, 11(1), pp. 26–35, 2017. DOI: 10.1049/iet-spr.2016.00101

25. S. Kumar, A. Kumar, "A brief review on antlion optimization algorithm," in *2018 International Conference on Advances in Computing, Communication Control. and Networking (ICACCCN)*, pp. 236–240, 2018.

26. S. Mirjalili, "The ant lion optimizer," *Advances in Engineering Software*, 83, pp. 80–98, 2015.

27. Wang, S. Li, "An improved grey wolf optimizer based on differential evolution and elimination mechanism," *Scientific Reports*, 9(1), pp. 71–81, 2019.

28. S. Shrivastava, D. K. Upadhyay, O. P. Goswami, "Optimal design of fractional-order low-pass filter using L2-method," *2021 International Conference on Communication, Control and Information Sciences (ICCISc)*, pp. 1–5, 2021. DOI: 10.1109/ICCISc52257.2021.9484936

29. H. Faris, M. A. Hassonah, A. M. Al-Zoubi, S. M. Mirjalili, I. Aljarah, "A multi-verse optimizer approach for feature selection and optimizing SVM parameters based on a robust system architecture," *Neural Computing and Applications*, 30(8), pp. 2355–2369, 2018. DOI: 10.1007/s00521-016-2818-2

30. S. Mirjalili, S. M. Mirjalili, A. Hatamlou, "Multi-verse optimizer: A nature-inspired algorithm for global optimization," *Neural Computing and Applications*, 27(2), pp. 495–513, 2016.

6 Applications of Reinforcement Learning in the Physical Layer of Wireless Communication

Dhiraj Shrivastava, Akash Gupta and Bharat Verma

6.1 INTRODUCTION

A new wireless communication paradigm, the sixth generation (6G) framework, with full Artificial Intelligence (AI) support, is planned for deployment by the end of this decade (1). The beyond 5G(B5G) communication systems will be an amalgamation of key technologies from the past (e.g., smaller cells, larger antenna arrays, massive MIMO, higher spectrum) and new technologies (reconfigurable intelligent surfaces, Machine Learning (ML) techniques). The development of the new standard is driven by the target area of application, e.g., massive machine-type traffic, mobile broadband and mission-critical applications. The B5G is envisioned as an intelligent wireless communication network that connects people and things.

For the 6G era, the four main drivers to the corresponding challenges that are emerging are (2): durability of frameworks at the centre of society, sustainability through the effectiveness of mobile technology, speed-up digitization to ease and enhance the lives of people and uninterrupted communication meets demands of accelerating communication to anywhere, anytime and anything. To meet these challenges, 6G must continue pushing technological boundaries beyond 5G, focusing on essential services, interactive communication systems and the omnipresent Internet of Things (IoT). Furthermore, additional capability dimensions should be investigated, such as integrating computation services and providing functionality beyond communication.

The vision for the B5G communication network includes providing ultra-high bit rates in the range of Tbps which will be available at particular sites. Further, the network will offer services with higher range, spectral and energy efficiency (3). Furthermore, latency is expected in ultra-long-range communication with less than 1 ms (4). The most exciting feature of 6G is that AI fully supports it for autonomous systems. Video traffic will likely outnumber other data traffic types in 6G communications. The THz band, AI, Optical Wireless Communications (OWC), 3D

DOI: 10.1201/9781003452645-6

networking, Unmanned Aerial Vehicles (UAV), IRS and wireless power transmission are the most significant technologies that will power the upcoming B5G communication technology.

The B5G wireless communication system targets more on the Quality of Service (QoS) than higher data rates. The QoS of a wireless system is mainly affected by the channel condition; interference caused due to users and overlying cells and handover impairments. The traditional approach of designing and analysing a communication system does not support future complex wireless networks with regard to operation and optimization. Reinforcement Learning (RL) is emerging as one of the prominent technology that can aid network designers to cater to the demands of exponentially increasing data users. This chapter deals with the need for integration of RL in the existing communication system and also focuses on the various RL schemes proposed in the field of wireless communication.

6.1.1 REINFORCEMENT LEARNING

RL is a branch of ML that relies on continuous feedback mechanisms. In RL, an agent learns from the environment by acting and observing the results. Feedback is returned from the environment to the agent, and the agent's actions are evaluated. Agents receive positive feedback or rewards for each good action and negative feedback or punishment for each bad action.

RL algorithms are based on model-free and model-based learning strategies (5).

Model-free RL algorithm: This strategy is founded on learning via trial and error.

Model-based RL algorithm: This technique relies on learning by planning because of its fast convergence and simplicity.

Deep Reinforcement Learning (DRL), based on iterative experience in a training process, depends on the agent completing a task determined by the reward function. The DRL agent learns to model actions and results without having to model the underlying physical process that produces the result. DRL consists of two algorithms, policy gradients and Deep Q Learning (DQL). DRL has a variety of uses in networking and communication, including traffic routing, resource scheduling, caching and offloading, privacy and connectivity preservation, network access and rate control and data gathering (6). Figure 6.2 shows the taxonomy of the RL algorithm.

6.2 HANDOVER

The handover mechanism is used in mobile communication, where a connected data session or cellular call is transferred from one cell site to another without disconnection. In cellular services, handover is one of the critical characteristics as the users' QoS must not be compromised while it migrates from one cell to another. The handover handles calls and maintains the data session, regardless of whether the user moves between the cells (horizontal handover) or migrates from one network to another (vertical handover). Heterogeneous networks (Hetnets) are one of the important paradigms in 5G networks, and thus, migration of users from one network to another will be more frequent. The user terminal will experience a degraded QoS if the timely handover is not happening or false handovers are increased. Therefore, to ensure no QoS is affected and unnecessary handovers are avoided, the handover must occur accurately by correctly triggering the handover decision considering all the necessary parameters (7).

In the literature, multiple works have focused on optimizing the handover scheme to increase overall throughput and reduce false handover. In the available literature, various handover mechanisms have been proposed, such as alternative approximate methods, hysteresis-based handoff algorithms, local averaging and a drop timer algorithm (8), a local averaging technique (9), Received Signal Strength Indicator (RSSI) method, 3D positioning algorithm (10), Vertical Handover (VHO) scheme (11), etc. The schemes proposed in the literature have their advantages. Still, they can be made more robust and flexible by implementing RL. Researchers across the globe are exploring the applicability of RL schemes in the handover mechanism.

The authors in (12) proposed RL-based optimization of the handover technique. The handover between the base station was controlled using the centralized RL agent. The authors analysed the performance for various deployments of propagation environments and compared the results with existing literature. The result indicates a link beam performance gain of 0.3 to 0.7 dB for realistic propagation environments. The authors (13) recommend a certified DRL-based Adaptive Handover Mechanism for VLC in a hybrid 6G network architecture to regulate an ultra-dense deployment of VLC Access Points (APs), which offers various issues with user mobility and lowering the interruption time by handover. The authors have evaluated the result with the traditional RL algorithm. The result indicates that the average downlink data rate is up to 48% better than those with conventional RL algorithms. This algorithm also performs better than Deep Queue Network (DQN), State Action Reward State Action (SARSA) and Q-learning algorithms by 8%, 13% and 13%, respectively. The authors in (14) proposed a two-layer framework DRL-based optimization of the handover technique. The first split User Equipment (UE) into clusters with different mobility patterns and then used the RL framework to obtain the optimal controller for each UE with the same cluster. The authors analysed the result and compared it with existing literature. The result showed that the Upper Confidence Bandit (UCB) algorithm achieves better throughput and lower HO rate after the same learning time, which is already 80% better than the traditional 3GPP method.

The authors in (15) proposed a DRL-based optimization adaptation for handover. The millimetre-wave network paradigm shifts for leveraging time-consecutive camera images in handover were controlled using the DRL method. The author analysed the performance and compared the result with existing literature. The authors predict the cumulative sum of future data rates, fix handover times based on predicted values based on DRLs, and state information based on experimentally obtained camera images. This state expansion helps to predict the interrupted data rate degradation, which is about 500 ms before the degradation occurs. The average time for the data rates to be discovered is about two more minutes. The result indicates that the proposed state space enhances performance gain. The authors in (16) proposed a DRL-based handover mechanism for High-Speed Train (HST) applications. In HST, due to handover, the degradation problem was handled using the Real-Time Transport Control Protocol (RTCP) method based on the DRL method. The author analysed the performance of degradation problems and compared the results with existing literature. The simulation result depicted that with the RTCP approach, the throughput increases in a static and high-speed environment by around 2% and 67%, respectively, compared with all TCP mechanisms. The authors in (21) proposed a Double Deep Q-learning (DDQN)-based handover algorithm for vehicle-to-network communications and compared the simulated result with some state-of-the-art works available, which is shown in Table no. 6.1.

6.3   INDOOR LOCALIZATION

Indoor device localization has been extensively investigated in the past few decades, mainly in industrial settings, wireless sensor networks and robotics. In the literature, various indoor localization mechanism has been proposed, like the simplified quaternion-based orientation estimation algorithm, novel quaternion Kalman filtering (22), Support Vector Machine (SVM), K-Nearest algorithm (KNN) (23), Proximity Detection algorithm (24), simultaneous localization and mapping (25), barcode-based navigation (26), Hidden Markov model (27), Recursive Bayesian Filtering (RBF), PF-based map matching algorithm (28), WKNN/PDR map matching algorithm (29) was used. The RL-based indoor localization aims to provide faster and more accurate results.

The authors in (30) proposed RL-based optimization for the indoor localization technique. The direction of each step and calculation of the location after moving was controlled using the RL agent. The authors used a novel direction calibration algorithm, analysed the localization performance with average error for Direction calibration with RL (RLDC) and Pedestrian Dead Rocking (PDR), and found that the performance was improved by approximately 67% by RLDC. In (31), the authors proposed an RL-based adaptation for indoor localization. The prediction model was used to predict the user's location using a prediction model based on ML techniques. The authors analysed ML-based predictive models and compared them with different models and found that the accuracy was ML at 97.83%, deep learning (DP) at 97.67%, AI at 97.33% and RL at 98.50%. In (32), the authors proposed RL optimization for the indoor space recognition method. The perceptual model detected the indoor environment due to transient phenomena such as obstacles and fading. The authors analysed the perceptual model with the deep learning method and found that accuracy was improved to 90% compared to the traditional approach. In (33), the authors proposed a DRL adaptation for indoor navigation systems. Using a deep RL algorithm, this algorithm converted visual data into a 2D environment. The authors analysed the performance of deep learning with the 2D bar code navigation method and SLAM. They found that the navigation time was improved by a minimum and maximum of 92 seconds and 129 seconds.

6.4   HETEROGENEOUS NETWORKS (HETNETS)

The upcoming network generation, known as HetNet, will be the one to unify telecom infrastructure below a single network's wing. Users will be able to take

TABLE 6.1
Comparison of Different State-of-the-Art Methods with the DDQN HO Algorithm (21)

Algorithm	A3 RSRP (17)	Proposed DDQN HO	Q-learning-based HO Parameter Optimization (18)	Multiuser DRL (14)	FL Proactive HO Trigger (19)	Two-tier Proactive HO Optimization (20)
Methodology	Static control scheme	RL	RL	Unsupervised learning RL	Supervised learning	Supervised learning RL
Key idea	Fixed parameter-based HO trigger to strongest BS	HO decision optimization	Parameter optimization for A3-based HO	Mobility-based UE clustering, cluster-level optimal HO policies	SNR prediction for HO trigger	RSSI prediction for HO trigger, RL-based HO decision optimization
Learning technique setups	–	Deep Q-learning	Traditional Q-learning	K-means clustering, A3C policy gradient	Feedforward ANN-based prediction, Federated learning	RNN (prediction), HMM (HO decision)
Input parameter	RSRP	RSRP	Throughput, Packet delay, HO frequency	UE mobility, RSRQ, Throughput, HO frequency	SNR	RSSI
Execution agent	Core Network	Core Network	Core Network	Core Network	Distributed	UE (prediction), Core network (HO frequency)
Impact on network architecture	No	Low	Moderate	High	Very High	High
Communication overhead	Very low	Moderate	Low	Moderate	High	Moderate

advantage of continuous, unlimited internet working thanks to HetNet's interconnection of small cell infrastructure, macro infrastructure or unregulated technologies like Wi-Fi. Wide-area HetNets will include every component of a wireless network in any deployment scenario, including indoor and outdoor settings, office buildings, residential structures and underground spaces.

In the literature (34–44), researchers have proposed and analysed HetNets models without utilizing RL/ ML algorithms. However, with the goal of providing more adaptability to the existing systems, researchers across the globe are exploring the application of RL in optimizing the HetNets.

In (45), the authors proposed optimizing the HetNet scheme using RL. Co-channel interference in femtocells was managed using a multi-agent RL-based re-source management method. The authors analysed and compared the result with existing literature. The Macro Bank Station (MBS) and Femto Base Station (FBS) adjust the transmitting power and select the appropriate power for the system efficiency to improve power efficiency. The result indicates that with Deep Q Network (DQN), the energy efficiency is two times higher than that of Q-learning AA. In (46), the authors proposed the RL-based optimization of the HetNet method. In the dense small cell and heterogeneous network, frequent handover, handover failure and mobility were controlled using an RL-based approach. The authors analysed and compared the result with the existing literature and found that using an autonomous mobility management control approach reduced the handover by about 20%, latency by about 58%, handover failure rate by approximately near zero and throughput increased by about 12%. The authors (47) proposed a DRL-based optimization technique of HetNet. The unpredicted traffic in 5G, due to the high mobility of users, was controlled by a DRL-based intelligent Time Division Duplex (TDD) scheduling algorithm. The authors analysed and compared the results with the existing literature and found that the Deep Reinforcement Learning-based Time Division Duplex (TDD) algorithm reduced packet loss rates by about 30% and increased throughput by about 1.6 times compared to the traditional method. The authors in (48) proposed HetNet's DRL-based optimization technology. In a HetNet, handover and power allocation were controlled using a Multi-Agent Reinforcement Learning (MARL) algorithm. The authors analysed performance through centralized training with a decentralized execution framework, compared existing literature to each UE's training policies and found results indicating improved performance with increased throughput and reduced HO frequency. The authors in (49) proposed a DRL-based optimization technique of HetNet. When the number of users dynamically changed, the power allocation was controlled using the padding. The authors analysed the power allocation and capacity performance and compared the result with the existing literature. When the scenario contains one Microcell Base Station (MBS), seven small Cell Base Stations (SBS) and 20 users compared to the WMMSE algorithm and max-RSS algorithm, the result indicates that the total system capacity improved by 18.4% and 50%, respectively. The authors in (50) proposed a DRL-based optimization technique of HetNet. Co-channel interference in multiple D2MD clusters coexisting with base stations (BS) and cellular users (CUs) was controlled using the multi-agent DRL-based DPRA algorithm. The authors evaluated the results by comparing the existing literature. The result indicates that the applied DPRA algorithm increases the average summation rate by three times compared with the traditional methods, so it also increases the robustness of the performance. The authors in (51) proposed HetNet DRL-based optimization technology. VR transmission requires bandwidth, and the Quality of Services (QoS) was controlled using a novel hybrid transmission mode selection based on online RL. The authors analysed and compared the results with the existing literature. The result indicates that the Win or Learn Fast Policy Hill Climbing (WOLF-PHC) and Nash Q-learning algorithms increase throughput by 2.5 times compared to traditional approaches. The result showed better system throughput for VR broadband services with moderate resource cost than game theory broadcast plans in 5G HetNets. The authors in (52) proposed HetNet DRL-based optimization technology, HetNets and so on, radio resource allocation and management (RRAM), were controlled using DRL techniques. The authors analysed and compared the existing literature and found that efficient performance improved complex wireless optimization problems, including RRAM problems. The authors in (53) proposed a DRL-based optimization technology for the HetNets. In HetNet, energy efficiency maximization in small cell base stations was handled by multi-agent DQL. The authors analysed and compared the results with the existing literature. The result indicates that with the Deep Queue Learning (DQL) approach, the energy efficiency is increased by about 17% compared to the traditional method.

6.5   RESOURCE ALLOCATION

Optimum power allocation improves the efficiency of wireless systems. In the literature (54–57), conventional mathematical framework-based optimum power allocation schemes are presented and analysed. The RL-based optimization will further improve the existing schemes to allocate the required power to individual users. The authors in (58) proposed a DRL-based optimization method for power allocation. In Multiple Target Tracking (MTT), radar power allocation was controlled using a DRL-based algorithm. The authors analysed the power allocation performance, compared the results with the existing literature and found that the power allocation performance is better than the traditional method.

The authors in (59) proposed a DRL-based optimization method of power allocation. In Cell-Free (CF) massive Multiple-Input Multiple-Output (MIMO) systems, the power allocation was controlled using a certified DRL algorithm. The authors analysed the performance of power allocation and spectral efficiency and compared the results with existing literature. The result indicates that the DRL-based method achieved at least 33% higher sum-SE than the WMMSE algorithm and that the execution time of the DRL method is 0.1In (60), the authors proposed a DRL-based optimization technique of power allocation. Power allocation of mmWave HSR systems was controlled using a certified Multi-Agent Deep Deterministic Policy Gradient (MADDPG) approach. The authors analysed the power allocation performance and compared the results with existing literature. The result indicates that the performance of the DRL-based MADDPG method is increased to about 77% compared to the conventional method. In (48), the authors proposed a DRL-based optimization technique of power allocation. To maximize the sum rate in a multi-user wireless cellular network, dynamic power was controlled using a multi-agent DRL framework. The authors analysed the sum rate performance, compared the results with the existing literature and found that the performance of the sum rate and robustness of the DRL algorithm-based system was higher than the traditional method.

The need for radio frequency spectrum for numerous applications, including high-speed data transport and communication, led to the allotment of the spectrum as wireless telecommunications technologies began to converge. Multiple schemes for efficient spectrum allocation are available in the literature, such as centralized resource allocation techniques, (61), fractional frequency reuse approach (62), resource allocation strategies (Power Control) (63), a heuristic algorithm for spectral efficiency, fixed margin algorithm, fixed power margin (64), full-spectrum reuse strategy (65) and Frequency Modulated Continuous Wave (FMCW) radar model (66). An RL-based mathematical framework for U2X communication (67) was used.

The authors in (68) proposed a DRL-based optimization method of spectrum allocation. In multi-level HetNets and device-to-device communication, spectrum allocation and throughput were controlled using a multi-agent Q learning-based approach. The authors analysed resource allocation and compared the results with existing literature. The result indicates that with Autonomous Spectrum Allocation (ASA) algorithm, the spectral efficiency and throughput performance are increased to about 23% and 19% compared to the conventional method. The authors (69) proposed a DRL-based optimization method of spectral allocation. In device-to-device (D2D), the communication infrastructure cellular network spectrum efficiency was controlled using a decentralized multi-agent deep RL. The authors analysed

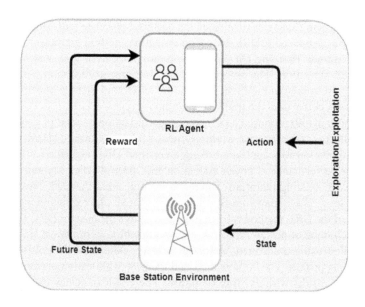

FIGURE 6.1   Reinforcement learning

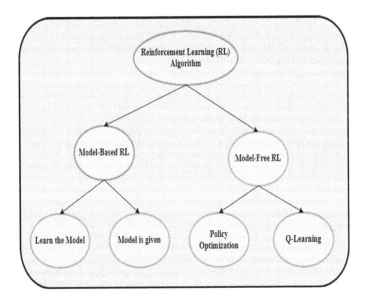

FIGURE 6.2   Taxonomy of reinforcement learning algorithm

the spectrum allocation performance with existing literature. The result indicates that the transmission quality of CUEs significantly improves the sum rate of D2D links and has better convergence than other traditional approaches. The authors in (70) proposed a DRL-based optimization technique of spectral allocation. In a large deployment and dynamic environment access and backhaul (IAB), network spectrum allocation was controlled using a deep reinforcement learning (DRL) approach. The authors analysed the performance gain and compared the result with existing literature. The result indicates that the actor-critic spectrum allocation (ACSA)-based algorithm increases the average summation rate to 20%, which is better than the traditional full-spectrum reuse strategy. The authors in (71) proposed a DRL-based optimization technique of spectral allocation. In automotive radar, mutual interference was controlled using an R-L-based algorithm. The authors analysed the result using the Q learning training algorithm and compared the result with existing literature. The result indicates that the RL-based trained Q-networks achieve around 10% success rate improvement over the myopic policy.

BIBLIOGRAPHY

1. M. Z. Chowdhury, M. Shahjalal, S. Ahmed, and Y. M. Jang, "6G wireless communication systems: Applications, requirements, technologies, challenges, and research directions," *IEEE Open J. Commun. Soc.*, 1(June), pp. 957–975, 2020.
2. G. Wikström et al., "6G - Connecting a cyber-physical world," no. February, 2022, Available: https://www.ericsson.com/en/reports-and-papers/white-papers/a-research-outlook-towards-6g.
3. K. David, and H. Berndt, "6G vision and requirements," *IEEE Veh. Technol. Mag.*, 13(July), pp. 72–80, 2018.

4. F. Tariq, M. R. A. Khandaker, K. K. Wong, M. A. Imran, M. Bennis, and M. Debbah, "A speculative study on 6G," *IEEE Wirel. Commun.*, 27(4), pp. 118–125, 2020.

5. B. Smith, R. Abay, J. Abbey, S. Balage, M. Brown, and R. Boyce, "Propulsionless planar phasing of multiple satellites using deep reinforcement learning," *Adv. Space Res.*, 67(11), pp. 3667–3682, 2021.

6. N. C. Luong, D. T. Hoang, S. Gong, D. Niyato, P. Wang, Y. C. Liang, and D. I. Kim, "Applications of deep reinforcement learning in communications and networking: A survey," *IEEE Commun. Surv. Tutor.*, 21(4), pp. 3133–3174, 2019.

7. P. Mahajan, and Zaheeruddin, "Review paper on optimization of handover parameter in heterogeneous networks," *3rd Int. Conf. Innov. Appl. Comput. Intell. Power, Energy Control. with their Impact Humanit. CIPECH*, pp. 111–115, 2018.

8. A. E. Leu, and B. L. Mark, "An efficient timer-based hard handoff algorithm for cellular networks," *IEEE Wirel. Commun. Netw. Conf. WCNC*, 2(C), pp. 1207–1212, 2003.

9. A. E. Leu, and B. L. Mark, "Modeling and analysis of fast handoff algorithms for microcellular networks," *Proc. IEEE Comput. Soc. Annu. Int. Symp. Model. Anal. Simul. Comput. Telecommun. Syst. MASCOTS*, 2002, pp. 321–328, 2002.

10. Y. Almadani et al., "Visible light communications for industrial applications—Challenges and potentials," *Electron.*, 9(12), pp. 1–38, 2020.

11. X. Bao, W. Adjardjah, A. A. Okine, W. Zhang, and J. Dai, "A QoE-maximization-based vertical handover scheme for VLC heterogeneous networks," *EURASIP J. Wirel. Commun. Netw.*, 2018(1), pp. 1–12, 2018.

12. V. Yajnanarayana, H. Ryden, and L. Hevizi, "5G handover using reinforcement learning," *2020 IEEE 3rd 5G World Forum, 5GWF 2020 - Conf. Proc.*, pp. 349–354, 2020.

13. L. Wang, D. Han, M. Zhang, D. Wang, and Z. Zhang, "Deep reinforcement learning-based adaptive handover mechanism for VLC in a hybrid 6G network architecture," *IEEE Access*, 9, pp. 87241–87250, 2021.

14. Z. Wang, L. Li, Y. Xu, H. Tian, and S. Cui, "Handover optimization via Asyn-chronous multi-user deep reinforcement learning," *6 IEEE Int. Conf. Commun.*, 2018(6), pp. 4296–4307, 2018.

15. Y. Koda, K. Nakashima, K. Yamamoto, T. Nishio, and M. Morikura, "Handover management for mmWave networks with proactive performance prediction using camera images and deep reinforcement learning," *IEEE Trans. Cogn. Commun. Netw.*, 6(2), pp. 802–816, 2020.

16. J. Xu, B. Ai, L. Wu, and L. Chen, "Handover-aware cross-layer aided TCP with deep reinforcement learning for high-speed railway networks," *IEEE Netw. Lett.*, 3(1), pp. 31–35, 2020.

17. K. Dimou et al., "Handover within 3GPP LTE: Design principles and performance," *2009 IEEE 70th Veh. Technol. Conf. Fall*, Anchorage, AK, pp. 1–5, 2009.

18. A. Abdelmohsen, M. Abdelwahab, M. Adel, M. Saeed Darweesh, and H. Mostafa, "LTE handover parameters optimization using Q-learning technique." *Midwest Symp. Circuits Syst.*, 2018, pp. 194–197, 2019.

19. K. Qi, T. Liu, and C. Yang, "Federated learning based proactive handover in millimeter-wave vehicular networks," *Intl. Conf. on Signal Process. Proc.*, ICSP, December, 2020, pp. 401–406, 2020.

20. N. Aljeri, and A. Boukerche, "A two-tier machine learning-based handover management scheme for intelligent vehicular networks," *Ad Hoc Netw.*, 94(v), 2019.

21. K. Tan, D. Bremner, J. Le Kernec, Y. Sambo, L. Zhang, and M. A. Imran, "Intelligent handover algorithm for vehicle-to-network communications with double-deep Q-learning," *IEEE Trans. Veh. Technol.*, 71(7), pp. 7848–7862, July 2022.

22. H. Bao, and W. C. Wong, "An indoor dead-reckoning algorithm with map matching," *2013 9th Int. Wirel. Commun. Mob. Comput. Conf.* IWCMC, 2013, pp. 1534–1539, 2013.

23. C. Jiang, H. Zhang, Y. Ren, Z. Han, K.-C. Chen, and L. Hanzo, "Machine learning paradigms for next-generation wireless networks," pp. 2–9, 2016.

24. S. A. Cheraghi, V. Namboodiri, and L. Walker, "GuideBeacon: Beacon-based indoor wayfinding for the blind, visually impaired, and disoriented," *2017 IEEE Int. Conf. Pervasive Comput. Commun.* (PerCom) 2017, pp. 121–130, 2017.

25. C. Röger, and S. Timpf, "Indoor mapping for human navigation – A low-cost SLAM solution," *GI Forum*, 6(1), pp. 152–161, 2018.

26. L. George, and A. Mazel, "Humanoid robot indoor navigation based on 2D bar codes: Application to the NAO robot," *IEEE-RAS Int. Conf. Humanoid Robot.*, 2015(February), pp. 329–335, 2015.

27. S. Lamy-Perbal, N. Guenard, M. Boukallel, and A. Landragin-Frassati, "A HMM map-matching approach enhancing indoor positioning performances of an inertial measurement system," *2015 Int. Conf. Indoor Position. Indoor Navig.* IPIN, 2015(October), pp. 13–16, 2015.

28. M. T. Koroglu, and A. Yilmaz, "Pedestrian inertial navigation with building floor plans for indoor environments via non-recursive Bayesian filtering," *Proc. IEEE Sens.*, 2017, pp. 1–3, 2017.

29. F. Peng, and J. Zhai, "A WKNN/PDR/Map-matching integrated indoor location method," *Proc. IEEE Int. Conf. Softw. Eng. Serv. Sci.* ICSESS, 2018, pp. 182–185, 2019.

30. Q. Li, X. Liao, and Z. Gao, "An enhanced direction calibration based on reinforcement learning for indoor localization system," *IEEE Wirel. Commun. Netw. Conf. WCNC*, 2020(May), 2020.

31. J. C. D. Cruz, and T. M. Amado, "Development of machine learning-based predictive models for wireless indoor localization application with feature ranking via recursive feature elimination algorithm," *ICSPCC 2020 IEEE Int. Conf. Signal Process. Commun. Comput. Proc.*, pp. 1–6, 2020.

32. Efficient indoor localization via reinforcement learning Dimitris Milioris Nokia Bell Labs, 91620 Nozay, France Massachusetts Institute of Technology, 77 Massachusetts Avenue, MA 02139, USA, pp. 8350–8354, 2019.

33. V. A. Bakale, Y. Kumar, V. C. Roodagi, Y. N. Kulkarni, M. S. Patil, and S. Chickerur, "Indoor navigation with deep reinforcement learning," *Proc. 5th Int. Conf. Inven. Comput. Technol.* ICICT 2020, pp. 660–665, 2020.

34. V. N. Ha, and L. B. Le, "Fair resource allocation for OFDMA femtocell networks with macrocell protection," *IEEE Trans. Veh. Technol.*, 63(3), pp. 1388–1401, 2014.

35. Y. Zhong, H. Wang, and H. Lv, "A cognitive wireless networks access selection algorithm based on MADM," *Ad Hoc Netw.*, 109(June), p. 102286, 2020.

36. S. M. A. Zaidi, M. Manalastas, H. Farooq, and A. Imran, "Mobility management in emerging ultra-dense cellular networks: A survey, outlook, and future research directions," *IEEE Access*, 8, pp. 183505–183533, 2020.

37. S. Song, Y. Chang, H. Xu, D. Zheng, and D. Yang, "Energy efficiency model based on stochastic geometry in dynamic TDD cellular networks," *2014 IEEE Int. Conf. Commun. Work.* ICC, 2014, pp. 889–894, 2014.

38. T. Goyal, and S. Kaushal, "Handover optimization scheme for LTE-advance networks based on AHP-TOPSIS and Q-learning," *Comput. Commun.*, 133, pp. 67–76, 2019.

39. Q. Shi, M. Razaviyayn, Z. Q. Luo, and C. He, "An iteratively weighted MMSE approach to distributed sum-utility maximization for a MIMO interfering broadcast channel," *IEEE Trans. Signal Process.*, 59(9), pp. 4331–4340, 2011.

40. K. Shen, and W. Yu, "Fractional programming for communication systems - Part I: Power control and beamforming," *IEEE Trans. Signal Process.*, 66(10), pp. 2616–2630, 2018.

41. C. Song, and Y. Jeon, "Weighted MMSE precoder designs for sum-utility maximization in multi-user SWIPT network-MIMO with per-BS power constraints," *IEEE Trans. Veh. Technol.*, 67(3), pp. 2809–2813, 2018.

42. J. Zheng, Y. Cai, X. Chen, R. Li, and H. Zhang, "Optimal base station sleeping in green cellular networks: A distributed cooperative framework based on game theory," *IEEE Trans. Wirel. Commun.*, 14(8), pp. 4391–4406, 2015.

43. J. J. Gimenez et al., "5G new radio for terrestrial broadcast: A forward-looking approach for NR-MBMS," *IEEE Trans. Broadcast.*, 65(2), pp. 356–368, 2019.

44. S. Zhang, N. Zhang, S. Zhou, J. Gong, Z. Niu, and X. Shen, "Energy-aware traffic offloading for green heterogeneous networks," *IEEE J. Sel. Areas Commun.*, 34(5), pp. 1116–1129, 2016.

45. Q. Su, B. Li, C. Wang, C. Qin, and W. Wang, "A power allocation scheme based on deep reinforcement learning in HetNets," *2020 Int. Conf. Comput. Netw. Commun.* ICNC, 2020, pp. 245–250, 2020.

46. Q. Liu, C. F. Kwong, S. Zhou, T. Ye, L. Li, and S. P. Ardakani, "Autonomous mobility management for 5G ultra-dense HetNets via reinforcement learning with tile coding function approximation," *IEEE Access.*, 9, pp. 97942–97952, 2021.

47. F. Tang, Y. Zhou, and N. Kato, "Deep reinforcement learning for dynamic uplink/downlink resource allocation in high Mobility 5G HetNet," *IEEE J. Sel. Areas Commun.*, 38(12), pp. 2773–2782, 2020.

48. D. Guo, L. Tang, X. Zhang, and Y. C. Liang, "Joint optimization of handover control and power allocation based on multi-agent deep reinforcement learning," *IEEE Trans. Veh. Technol.*, 69(11), pp. 13124–13138, 2020.

49. Y. Chen, and H. Zhang, "Power allocation based on deep reinforcement learning in HetNets with varying user activity," *2020 IEEE Glob. Commun. Conf. GLOBECOM 2020 - Proc.*, 2020.

50. Y. Xiao, J. Wu, and J. Liu, "Power allocation for device-to-multi-device enabled HetNets: A deep reinforcement learning approach," *2021 IEEE Glob. Commun. Conf. GLOBECOM 2021 - Proc.*, pp. 2–7, 2021.

51. L. Feng, Z. Yang, Y. Yang, X. Que, and K. Zhang, "Smart mode selection using online reinforcement learning for VR broadband broadcasting in D2D assisted 5G HetNets," *IEEE Trans. Broadcast.*, 66(2), pp. 600–611, 2020.

52. A. Alwarafy, M. Abdallah, B. S. Ciftler, A. Al-Fuqaha, and M. Hamdi, "The frontiers of deep reinforcement learning for resource management in future wireless HetNets: Techniques, challenges, and research directions," *IEEE Open J. Commun. Società*, 3(February), pp. 322–365, 2022.

53. B. Gu, Y. Wei, X. Liu, M. Song, and Z. Han, "Traffic offloading and power al- location for green hetnets using reinforcement learning method," *2019 IEEE Glob. Commun. Conf. GLOBECOM 2019 - Proc.*, 2019.

54. H. Godrich, A. P. Petropulu, and H. V. Poor, "Power allocation strategies for target localization in distributed multiple-radar architectures," *IEEE Trans. Signal Process.*, 59(7), pp. 3226–3240, 2011.

55. Y. S. Nasir, and D. Guo, "Multi-agent deep reinforcement learning for dynamic power allocation in wireless networks," *IEEE J. Sel. Areas Commun.*, 37(10), pp. 2239–2250, 2019.

56. J. Kim, G. Lee, and H. P. In, "Adaptive time-to-trigger scheme for optimizing LTE handover," *Int. J. Control Autom.*, 7(3), pp. 35–44, 2014.

57. C. Shen, and M. Van Der Schaar, "A learning approach to frequent handover mitigations in 3GPP mobility protocols," *IEEE Wirel. Commun. Netw. Conf.* WCNC, 2017.

58. Y. Shi, B. Jiu, J. Yan, H. Liu, and K. Li, "Data-driven simultaneous multibeam power allocation: When multiple targets tracking meets deep reinforcement learning," *IEEE Syst. J.*, 15(1), pp. 1264–1274, 2021.
59. Y. Zhao, I. G. Niemegeers, and S. M. H. De Groot, "Dynamic power allocation for cell-free massive MIMO: Deep reinforcement learning methods," *IEEE Access.*, 9, pp. 102953–102965, 2021.
60. J. Xu, and B. Ai, "Experience-driven power allocation using multi-agent deep reinforcement learning for millimeter-wave high-speed railway systems," *IEEE Trans. Intell. Transp. Syst.*, 23(6), pp. 5490–5500, 2022.
61. T. Peng, Q. Lu, H. Wang, S. Xu, and W. Wang, "Interference avoidance mechanisms in the hybrid cellular and device-to-device systems," *IEEE Int. Symp. Pers. Indoor Mob. Radio Commun.* PIMRC, pp. 617–621, 2009.
62. H. S. Chae, J. Gu, B. G. Choi, and M. Y. Chung, "Radio resource allocation scheme for device-to-device communication in cellular networks using fractional frequency reuse," *17th Asia-Pacific Conf. Commun.* APCC, 1(October), pp. 58–62, 2011.
63. G. J. Foschini, and Z. Miljanic, "A simple distributed autonomous power control algorithm and its convergence," *IEEE Trans. Veh. Technol.*, 42(4), pp. 641–646, 1993.
64. Y. Jiang, Q. Ling, and C. Zheng, "Social-aware device-to-device communications underlaying cellular networks," *J. China Univ. Posts Telecommun.*, 25(1), pp. 29–36, 2018.
65. B. Zhuang, D. Guo, E. Wei, and M. L. Honig, "Large-scale spectrum allocation for cellular networks via sparse optimization," *IEEE Trans. Signal Process.*, 66(20), pp. 5470–5483, 2018.
66. G. Kim, J. Mun, and J. Lee, "A peer-to-peer interference analysis for auto-motive chirp sequence radars," *IEEE Trans. Veh. Technol.*, 67(9), pp. 8110–8117, 2018.
67. S. Zhang, H. Zhang, and L. Song, "Beyond D2D: Full dimension UAV-to-everything communications in 6G," *IEEE Trans. Veh. Technol.*, 69(6), pp. 6592–6602, 2020.
68. K. Zia, N. Javed, M. N. Sial, S. Ahmed, A. A. Pirzada, and F. Pervez, "A distributed multi-agent RL-based autonomous spectrum allocation scheme in D2D enabled multi-tier HetNets," *IEEE Access*, 7(Figure 1), pp. 6733–6745, 2019.
69. Z. Li, and C. Guo, "Multi-agent deep reinforcement learning based spectrum allocation for D2D underlay communications," *IEEE Trans. Veh. Technol.*, 69(2), pp. 1828–1840, 2020.
70. W. Lei, Y. Ye, and M. Xiao, "Deep reinforcement learning-based spectrum Al-location in integrated access and backhaul networks," *IEEE Trans. Cogn. Commun. Netw.*, 6(3), pp. 970–979, 2020.
71. P. Liu, Y. Liu, T. Huang, Y. Lu, and X. Wang, "Decentralized automotive radar spectrum allocation to avoid mutual interference using reinforcement learning," *IEEE Trans. Aerosp. Electron. Syst.*, 57(1), pp. 190–205, 2021.

7 Computational Intelligence in MAC Layer Protocols of mmWave (5G and Beyond) V2X Communication

Aakash Jasper, Raghavendra Pal, Arun Prakash, Sara Paiva and Rajeev Tripathi

7.1 INTRODUCTION

Communication has played a vital role in road safety and traffic management of vehicles – from simple traffic signals to advanced signals in the future, which will facilitate and regulate the movement of driverless cars and make the journey of the user safe and convenient. However, as the level of automation in a vehicle increases (or dependence for driving a car manually by a driver decreases), the vehicular communication system's Quality of Service (QoS) requirements becomes more challenging to fulfil. The QoS for a given communication application requires the service provider to ensure a certain level of performance. These performance parameters are data rate, latency, reliability, communication range etc. Legacy communication generations, such as 2G and 3G, used controls and data processing at the central core network which resulted in a burden on the central core and non-scalability of the vehicular network. Also, latency is very high since the decision is being taken away from the vehicle at the central core. Therefore, in 4G, 5G and beyond the architecture has changed and decisions are taken at the edge of the network (Base Station (BS)) which is closer to the user. This is the reason that the Medium Access Control (MAC) is required to be made smarter with algorithms that have low computational complexity to work in real-time scenarios. As the level of automation in vehicles increases, the constraint on the QoS performance parameters becomes more stringent. The Society of Automotive Engineers (SAE) International, has defined a scale that has set the level of automation in a vehicle

DOI: 10.1201/9781003452645-7

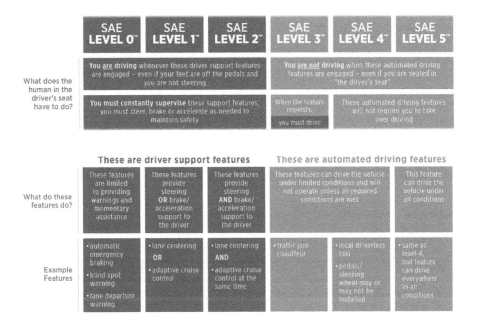

SAE LEVEL 0™	SAE LEVEL 1™	SAE LEVEL 2™	SAE LEVEL 3™	SAE LEVEL 4™	SAE LEVEL 5™

FIGURE 7.1 SAE levels of driving automation [1]

(Figure 7.1). The scale goes from Level 0 to Level 5 in the increasing order of vehicular automation. Level 0 indicates that the driving of the vehicle is fully manual and Level 5 stands for fully automatic driving. This means that the QoS requirement for vehicular communication at Level 0 will be very lenient and simple to fulfil since a human being is driving the vehicle manually while at Level 5 of the scale, there is no human involvement but all vehicular manoeuvres and control are being done by machines in the vehicle. A human driver can see the road, obstacles, traffic signals, pedestrians crossing the road, speed breakers, i.e., the entire traffic scenario around him. Considering the human driver as a complex highly advanced system, the input to this system is the vision of the road traffic scenario. The input goes through the eyes of the human driver (eyes can be considered here as the sensor), the situation is processed by the brain which gives signals to the hands and feet of the driver to drive the vehicle. However, a human being is not perfect and his driving vision is limited no matter how good a driver he/she is. As a result, we see that as the number of vehicles on the road is growing exponentially with time so is the number of road accidents which result in harm to the human beings involved and in the worst case even permanent damage or death. Not only road accidents but also since human vision is limited, he/she does not know the best route to travel, the upcoming road blockage, status of the parking etc., which results in heavy traffic jams that result in air pollution due to the fuel emissions of a vehicle. Figure 7.2 shows road deaths in different parts of the world as per the WHO Global Status Report on Road Safety 2018 [2]. On observing the graph in Figure 7.2, with time road deaths have increased in South-East Asia and Africa while in Europe and America road deaths

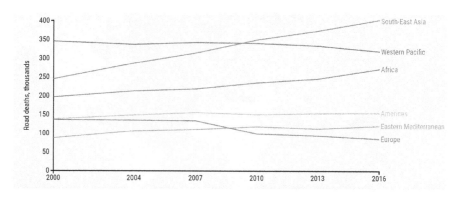

FIGURE 7.2   Road death statistics around the world from 2000 to 2016 [2]

have decreased or remained almost constant. One of the reasons for this is that the vehicles used in South-East Asia and Africa are basic and completely manual while the vehicles in America and Europe use vehicular communication systems such as (Dedicated Short-Range Communication (DSRC based on IEEE 802.11p protocol) [3] or 3GPP C-V2X [4] (the 4G Cellular Vehicle to everything communication system). These Intelligent Transportation Systems (ITS) use safety message packets in the form of Cooperative Awareness Messages (CAMs) or Basic Safety Messages (BSMs). These message packets contain basic information about the transmitting vehicle such as speed, direction of travel, acceleration etc. The packets are shared by a source vehicle to its surrounding vehicles for basic traffic management and safety purposes. The various ways for a vehicle to communicate with its surrounding can be categorized into the following communication links: Vehicle to Vehicle (V2V), Vehicle to Infrastructure (V2I) and Vehicle to Pedestrians (V2P).

Again, going back and observing the legacy cellular communication generations like 2G and 3G, many decisions such as handoff, channel management etc. were centralized, which resulted in huge data management at the central core of the network. However, this is non-ideal and impractical for vehicular communication where latency needs to be in the order of a few ms. As a result, the computation of heavy data was shifted closer to the user, i.e., BSs. This relieved the burden of heavy computation of large data at the central core. Mobile Edge Computing (MEC) [5] is a network architecture in which heavy user data are stored, processed and analysed near the user at the BS. The advantage of this is huge scalability of the network and reduction in latency of a given application since the distance of travel of data packet reduces significantly.

However, the current ITS like the DSRC (IEEE 802.11p) and 3GPP LTE C-V2X can provide very basic and simple safety messages for traffic safety. In the future, Connected Automated Vehicles (CAVs) will require very high QoS requirements which the current ITS both ad hoc (IEEE 802.11p) and cellular (LTE C-V2X) are unable to fulfil. For example, CAVs transfer raw sensor data (Radar, LIDAR, camera etc.) which may require data rate of up to 800 Mbps. IEEE 802.11p (max. data rate = 27 Mbps) and LTE C-V2X (max. data rate = 100 Mbps) are clearly unable to satisfy this QoS requirement. The reason for the abovementioned ITS to not provide such a high data rate requirement is that they operate at 20–40 MHz bandwidth at an operating frequency of sub-6 GHZ (5.9 GHz to be exact). The sub-6 GHz frequency band is already congested and heavily used by other communication and non-communication applications. The bandwidth at sub-6 GHz ranges is simply not enough to satisfy such high data rate requirement. The solution to this constraint on bandwidth is to go to higher frequency ranges (mmWaves – 30GHz – 300GHz) [6] which have large chunks of frequency bands that are unused.

The use of mmWave ranges seems to be the obvious logical choice. But as the frequency of an electromagnetic wave increases, its corresponding wavelength decreases. The small wavelength mmWaves encounter high path and penetration loss as compared to long wavelength radio waves. Even a human hand holding a mobile phone can severely attenuate the mmWave signal strength resulting in an outage. However, the small length of the mmWave also has the advantage that the antenna which will be used for operation will be of the size of the order of mm. As a result, highly directional antenna arrays can be designed to produce a very narrow beam with high SINR mmWave signal, as opposed to the omni-directional antennas used in sub-6 GHz V2X technologies.

The advanced use cases of vehicular communication like Vehicles Platooning, Advanced Driving, Extended Sensors and Remote Driving have very stringent requirements of QoS [7]. On studying the 3GPP technical specification [7], the strict communication QoS requirement of these advanced use cases can be observed. The reliability of all these four use cases must be in the range 90–99.999%. The reliability of remote driving must be strictly on or above 99.999%. This is because automatic manoeuvres of CAVs messages must contain no error. The maximum end-to-end latency for these use cases must be below 100 ms. Data rate may vary from 10 to 1000 Mbps depending on the advanced use case, especially extended sensors in which CAVs share raw sensor data at very high data rate [8].

To achieve these advanced vehicular QoS requirements IEEE is promoting IEEE 802.11bd (in place of IEEE 802.11p) and 3GPP is promoting 5G NR V2X (in place of 3GPP LTE C-V2X), both the new technologies capable of operating at mmWave frequency ranges.

7.2 COMPUTATIONAL INTELLIGENCE IN MAC LAYER OF VEHICULAR COMMUNICATION

From 2G to 5G and beyond the vehicular data processing and decision are taken closer to a user, i.e., moving from centralized architecture to de-centralized architecture. The modern CAVs must be capable of operating in both cellular and ad hoc architecture as in 3GPP 5G V2X communication (Mode 1 (V2I) and Mode 2 (V2V)). Figure 7.3 shows the 5G system architecture for V2X communication. In Figure 7.3, the PC5 link corresponds to the V2V communication link (link between two vehicles) and Uu is the V2I communication link (link between vehicle and infrastructure like BS or road side unit). V2V link can operate whenever the vehicles go into outage condition of the cellular architecture, which will happen often in mmWave

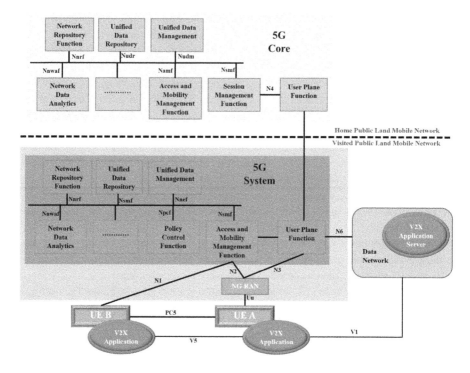

FIGURE 7.3   Architecture of 5G System for V2X communication using PC5 and Uu links – local breakout scenario (Roaming Case)

communication. Parallel working of Mode 1 and Mode 2 results in the overall increment of the mmWave network reliability and coverage.

The small range of mmWave coverage, use of highly directional antenna arrays (using beamforming techniques) and high path and penetration loss of mmWave make it a challenging task to design a reliable and scalable mmWave V2X communication system. Using physical devices operating at mmWave frequency ranges is not enough. Computational intelligence of smart algorithms and schemes must be used at the MAC layer level and MEC level to satisfy the QoS requirements of the upcoming smart CAVs.

The following are the three problem statements in mmWave V2X communication where computational intelligence of smart algorithms (schemes or Machine Learning (ML)) can be used to solve the problem:

1. Smart relay vehicle selection for data transfer in case of direct Line of Sight (LOS) blockage.
2. Fast initial access to the mmWave network by best beam selection.
3. Fair and efficient resource allocation to users of different categories (like Cellular user or Vehicular user, emergency user or non-emergency user message delivery).

7.2.1 Relay Vehicle Selection for Data Transfer and Perception Range Increment

To increase the perception range of a CAV and reliability of the communication link between source and destination CAV in a mmWave V2X network the use of an intermediate vehicle to act as a relay for data transfer becomes very crucial. Figure 7.4 shows Vehicular User Equipment (vUE) Mode 1 (V2I) and Mode 2 (V2V) relaying scenarios. The extended sensor use case requires that a CAV transfer its raw sensor data to its surrounding vehicles and receive sensor data from other vehicles. With the large sensor data received [9], the CAV acquires information on the speed, direction of travel, future intention, acceleration and other important telemetric data based on which it takes its future manoeuvre decisions. While travelling it may so happen that a V2V link may be occasionally blocked by some obstacle (e.g., Non-connected Vehicles (NV) etc.). In such a case also, an intermediate CAV may act as a relay to transfer data from source to destination.

Many research works are related to selecting a relay vehicle which increases the QoS of vehicular communication.

The general methodology by almost all researchers for vehicular relay selection involves the following three steps to be performed after the LOS link is blocked and before relay selection: (i) the source vUE sends a beacon signal to its surrounding vehicles, requesting them to act as a relay for data transfer to the destination, (ii) the surrounding vehicles sends an Acknowledgment/Negative-Acknowledgment (ACK/NACK) to the source vehicle depending upon whether they can/ cannot become a candidate for relay selection procedure and (iii) the last and the most important step, in which after receiving ACK/NACK signal from surrounding vehicles the source vUE constructs a set which contains the identity information (ids) of the candidate relay vehicles and applies a smart algorithm or scheme to select the best relay vehicle to maximize the QoS. In [10], based on the data traffic type (safety or infotainment) the characterization of the

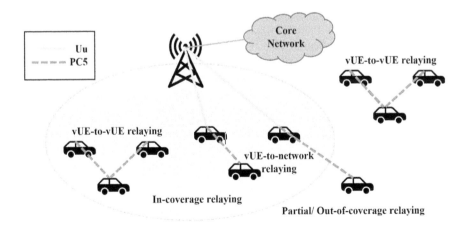

FIGURE 7.4 vUE relaying scenarios

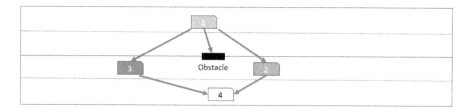

FIGURE 7.5 A simple scenario of vUE relay data transfer from source vUE (1) to destination vUE (4) when LOS (Line of Sight) is blocked by an obstacle

data transfer is done, i.e., higher priority given to safety data packets than the information data packets. This characterization is done using Analytical Hierarchical Process (AHP) which comes under Multiple Attribute Decision Making (MADM) problem solving. However, the disadvantage of this problem-solving scheme is that its decision table formation is subjective in nature. To reduce the subjectivity the authors have used shapely value in Coalition Game (CG) for the selection of the best relay vehicle in the candidate list. In [11], the authors use hybrid V2X RATs (Radio Access Technologies), mmWave for V2V and sub-6 GHz IEEE 802.11p for V2I. Figure 7.5 shows a simple scenario of relay data transfer from source vUE (1) to destination vUE (4) when an obstacle blocks the LoS. In this, for sensor data transfer from source vUE to destination vUE in case of blockage of direct LOS mmWave link, Diverted Path Algorithm (DPA) is used where the data are passed using a diverted path through a relay vUE. In case the DPA also fails then V2I link is used by the procedure of Backup Link Algorithm (BLA), as a result, the overall reliability of the V2X network increases. In [12], the relay selection is not done by the source vehicle but by the BS. The BS uses Opportunistic Relay Vehicle Selection (ORVS) scheme to select a relay for the source vUE data transmission. In ORVS, a local interference graph is created at the serving BS and is regularly updated by removing that relay which causes too many conflicts. In [13], Joint Optimization of the Single Hop Forward Progress and Latency (JOSHFPL) scheme is used for relay selection and reducing the latency parameter. The comparison of JOSHFPL scheme is done with other multi-hop data transfer schemes like Random with Forward Progress (RFP), Most Forward with fixed Radius (MFR) and Nearest with Forward Progress (NFP). In [14], mmWave 6G V2X system LOS Map is constructed and the following algorithms having certain computational complexity is compared: Congestion Game ($O(M^N N^{-N})$), Hungarian Game ($O(MN^2)$), First Come First Serve ($O(M)$) and Auction-based ($O(T_k M)$). In all these research works, first the system model, channel model and antenna model are designed and then an analysis of the relay selection scheme is done.

7.2.2 FAST INITIAL ACCESS TO THE MMWAVE V2X NETWORK

The sub-6 GHz V2X technologies (like DSRC and 3GPP LTE C-V2X) use omnidirectional antenna which makes the transmission and reception of signal simple. However, the mmWave V2X technologies (like 3GPP 5G NR V2X) use highly directional antenna arrays [15] since there will be high-power consumption [16] if omnidirectional antenna is used at mmWave frequency ranges. But directional data transmission makes the communication system more complex as the beams of transmitter and receiver antenna must be aligned. So smart beam search scheme needs to be done for beamforming. The initial access search [17] for the appropriate beam can be classified into the following categories: exhaustive search, hierarchical search, probabilistic search, meta-heuristic search, context-information search and ML search. In [18], a brute force algorithm is used where the entire beam space search is done in a sequential manner at both the source vUE and the serving RSU. The advantage of this scheme is that it searches for the best beam resulting in lower outage probability. However, the sequential search is a time-consuming process resulting in higher latency, and hence slow initial access. In [19], in this beam search method, the entire beam space set is divided into "n" number of steps. First, RSU scans the space using macro beams, after that the next step of the search is done in the best sector. This scheme has a lower initial access delay as compared to an exhaustive search but has the disadvantage of higher outage probability. In [20], statistical modelling of the mobility of the vehicular system and connection probability analysis is done before mmWave multilevel beamforming. This search method has both lower initial access delay and lower outage probability, but prior information of CAVs are needed and this scheme is not generic. In [21], the beam search scheme requires the prior context information of RSU/vUEs's position, status of network, QoS requirements, traffic flow etc., before the initial search. In [22], an ML search is done. This scheme also requires prior context information for dataset creation on which a multi-state Q-learning algorithm can be incorporated for best beam selection. This scheme is also prior information-dependent and not generic. In [23], a genetic algorithm which is a meta-heuristic search approach is used for initial access to mmWave network. These meta-heuristic beam search approaches are very good in performance since these algorithms intelligently use a smaller number of iterations.

7.2.3 EFFICIENT AND FAIR RESOURCE ALLOCATION FOR BETTER QoS IN MMWAVE V2X

Orthogonal Frequency Division Multiplexing (OFDM) is the modulation technique used in 3GPP 5G NR V2X. So, the CAVs in a V2X network are allocated time-frequency blocks called Resource Blocks (RBs) [24], after all the CAVs sense the availability of the channel during sensing time. In sensing time, the CAVs sense the availability of the resources and in selection time, based on various factors like priority, application type etc., RBs are allocated to the CAVs. The objective is to allocate the RBs efficiently and fairly to the CAVs for the best channel utilization and overall best QoS. The resource allocation can be centralized (like in Mode 1 of 3GPP 5G V2X) if the CAVs are in the coverage of the gNode-B (next-generation BS of a 5G network) or ad hoc/ distributed if the CAVs are out of coverage of the gNode-B (like in Mode 2 of 3GPP 5G V2X). Figure 7.6 shows the resource allocation of the CAVs. When there is no co-ordination among the CAVs, the same RB can be selected by multiple CAVs resulting in interference. The V2X resource allocation can be broadly classified into four categories [25]: interference aware, geo-based, time-frequency selection and clustering. In [26], Mixed Binary Integer Non-Linear Programming (MBINP) is used to minimize interference among vUEs and Cellular Users (CUs) with constraints on

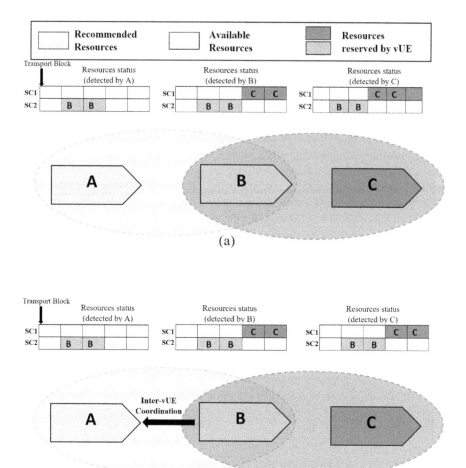

FIGURE 7.6   Advantage of inter vUE coordination to avoid the hidden terminal problem. (a) Without inter-vUE coordination (b) With inter-vUE coordination

QoS, total availability of power, interference threshold and minimum transmission rate. The disadvantages of this scheme are the limitation of transmission power of vUEs resulting in decreased SINR and no priority of signal is considered here which the 5G network promises. Short-Term Sensing-Resource Selection (STS-RS) [27] is also an interference-aware resource allocation mechanism. C-V2X uses this mechanism to reduce contention on resources. The disadvantage of this scheme is that as vUE move from one BS to another BS, it may rebuild a new resource selection map resulting in system inefficiency or failure. In [28], adaptive transmit power control algorithm is used which is also interference-aware resource allocation scheme. In this scheme, the overall QoS is improved by smartly assigning power to the vUEs for data transmission depending upon information gathered from beacon signals of other CAVs during channel sensing and the neighbour vehicle density. In [29, 30], geo-based scheduling scheme is used for resource allocation based on the channel-sensing information and location of the vUEs. The disadvantage of this scheme is that the vehicular system is highly dynamic, i.e., the channel and road condition (topology) vary frequently which may cause high latency and low data rate. Greedy cellular-based V2V link selection algorithm [31] is a time-frequency based resource selection scheme. Here, optimum resource selection is done at eNodeB (Enhanced BS of 4G network) to minimize latency. Cluster-Based Resource Block Sharing and Power Allocation (CROWN) [32] is used on a slow varying Channel State Information (CSI) V2X network with constraint on latency and reliability.

7.3   CONCLUSION AND FUTURE SCOPE

To reduce the latency of vehicular communication decision-making and data process-ing moving towards the edge of the network (BS and RSUs) are essential. Therefore, there is a need to develop smart MAC layer protocols. The mmWave V2X network has the frequent problem of vehicles going into the outage, so both V2I and V2V communication should be given equal importance. More automation of CAVs will result in less human involvement which will remove mistakes and inefficiency of human beings. This will result in fewer accidents, less pollution, best route selection etc. Some primary problems to realize V2X in the real world are best relay vehicle selection for sensor data transfer and perception range increment, fast initial access in mmWave V2X network and efficient and fair resource selection. The MAC layer protocols are made up of algorithms/ schemes whose performance can be evaluated based on their computational complexity. A smart algorithm, if it is to be used in real time must have less computational complexity as possible.

The algorithms used here are for the MAC layer. Smart decisions in long-term data analysis can be done at higher layers and central core networks. For example, central computer may collect and analyse the long-term vehicular travel history of a vUE and apply ML algorithms for prediction and overall journey efficiency (best travel route, less pollution, smart parking) etc. To make a vehicular system autono-mous is a hot topic and researchers are continuously working to come up with the best solution at both the lower and higher layers of data communication of vehicle.

In future, overall system analysis needs to be done considering both the MAC layer and higher layers. Studies on long-term vehicular data analysis in higher layers of the architecture and real-time decisions at MAC layers with their coordination also require thorough research. Top companies like QualComm, Samsung etc., are actively funding and working in modern vehicular communication.

REFERENCES

1. "SAE J3016 Levels of Driving Automation". sae.org/standards/content/j3016_202104.
2. "Global Status Report on Road Safety 2018". https://extranet.who.int/roadsafety/death-on-the-roads/#trends.

3. "IEEE 1609 – Family of Standards for Wireless Access in Vehicular Environments (WAVE)". U.S. Department of Transportation, Apr. 13, 2013.

4. "Study on LTE-Based V2X Services (v14.0.0, Release 14)". 3GPP, Sophia Antipolis, France, Rep. TR 36.885, Jul., 2016.

5. S. Wang, X. Zhang, Y. Zhang, L. Wang, J. Yang and W. Wang, "A Survey on Mobile Edge Networks: Convergence of Computing, Caching and Communications". *IEEE Access*, 5, pp. 6757–6779, 2017, doi: 10.1109/ACCESS.2017.2685434.

6. T. S. Rappaport, G. R. MacCartney, M. K. Samimi and S. Sun, "Wideband Millimeter-Wave Propagation Measurements and Channel Models for Future Wireless Communication System Design". *IEEE Transactions on Communications*, 63(9), pp. 3029–3056, Sept. 2015, doi: 10.1109/TCOMM.2015.2434384.

7. "Service Requirements for Enhanced V2X Scenarios (V16.2.0, Release 16)". 3GPP Standard TS 22.186, Jun., 2019.

8. M. Garcia, A. Galan, M. Boban, J. Gonzalvez, B. Perales, A. Kousaridas, "A Tutorial on 5G NR V2X Communications", *IEEE Communications Surveys and Tutorials*, 23(3), Third Quarter 2021.

9. A. D. Angelica, *Google's Self-Driving Car Gathers Nearly 1 GB/Sec* [Online], 2013. Available: http://www.kurzweilai.net/googles-self-driving-car-gathers-nearly-1-gbsec.

10. B. Fan, H. Tian, S. Zhu, Y. Chen and X. Zhu, "Traffic-Aware Relay Vehicle Selection in Millimetre-Wave Vehicle-to-Vehicle Communication", *IEEE Wireless Communications Letters*, 8(2), pp. 400–403, Apr. 2019.

11. R. Singh, S. Kumar and D. Saluja, "Reliability Improvement in Clustering-Based Vehicular Ad-Hoc Network", *IEEE Communications Letters*, 24(6), pp. 1351–1355, Jun. 2020.

12. J. Deng, O. Tirkkonen, R. Hollanti, T. Chen and N. Nikaein, "Resource Allocation and Interference Management for Opportunistic Relaying in Integrated mmWave/sub-6 GHz 5G Networks", *IEEE Communications Magazine*, 55(6), pp. 94–101, Jun. 2017.

13. Z. Li, L. Xiang, X. Ge, G. Mao and H. Chao, "Latency and Reliability of mmWave Multi-hop V2V Communications Under Relay Selections", *IEEE Transactions on Vehicular Technology*, 69(9), pp. 9807–9820, Sept. 2020.

14. F. Linsalata, S. Mura, M. Mizmizi, M. Magarini, P. Weng, M. N. Khormuji, A. Perotti and U. Spagnolini, "LoS-Map Construction for Proactive Relay of Opportunity Selection in 6G V2X Systems", arXiv, Cornell University, 15 Nov. 2021, doi: 10.48550/arXiv.2111.07804.

15. J. Wildman, P. H. J. Nardelli, M. Latva-Aho and S. Weber, "On the Joint Impact of Beamwidth and Orientation Error on Throughput in Directional Wireless Poisson Networks", *IEEE Transactions on Wireless Communications*, 13(12), pp. 7072–7085, Dec. 2014.

16. A. Yamamoto, K. Ogawa, T. Horimatsu, A. Kato and M. Fujise, "Path-Loss Prediction Models for Intervehicle Communication at 60 GHz". *IEEE Transactions on Vehicular Technology*, 57(1), pp. 65–78, Jan. 2008, doi: 10.1109/TVT.2007.901890.

17. I. Rasheed, "An Effective Approach for Initial Access in 5G-Millimeter Wave-Based Vehicle to Everything (V2X) Communication Using Improved Genetic Algorithm", *Physical Communication*, 52, p. 101619, 2022, ISSN 1874-4907, doi: 10.1016/j.phycom.2022.101619.

18. C. Liu, M. Li, S. V. Hanly, I. Collings and P. Whiting, "Millimetre Wave Beam Alignment: Large Deviations Analysis and Design Insights", *IEEE Journal on Selected Areas in Communications*, 35(7), pp. 1619–1631, Jul. 2017.

19. J. Qiao, X. Shen, J. Mark and Y. He, "MAC-Layer Concurrent Beamforming Protocol for Indoor Millimetre-Wave Networks", *IEEE Transactions on Vehicular Technology*, 64(1), pp. 327–338, Jan. 2015.

20. A. Abdelreheem, E. Mohamed and H. Esmaiel, "Location-Based Millimeter Wave Multi-level Beamforming Using Compressive Sensing", *IEEE Communications Letters*, 22(1), pp. 185–188, Jan. 2018.
21. F. Devoti, I. Filippini and A. Capone, "Facing the Millimeter-Wave Cell Discovery Challenge in 5G Networks With Context-Awareness", *IEEE Access*, 4, pp. 8019–8034, 2016.
22. Y.-J. Chen, W.-Y. Cheng and L.-C. Wang, "Learning-Assisted Beam Search for Indoor mmWave Networks", *2018 IEEE Wireless Communications and Networking Conference Workshops (WCNCW)*, pp. 320–325, 2018.
23. V. D. P. Souto, R. D. Souza, B. F. Uchôa-Filho and Y. Li, "A Novel Efficient Initial Access Method for 5G Millimeter Wave Communications Using Genetic Algorithm", *IEEE Transactions on Vehicular Technology*, 68(10), pp. 9908–9919, Oct. 2019.
24. "NR; Physical Channels and Modulation (V15.3.0, Release 15)". 3GPP Standard TS 38.211, Sep. 2018.
25. T. T. Thanh Le and S. Moh, "Comprehensive Survey of Radio Resource Allocation Schemes for 5G V2X Communications". *IEEE Access*, 9, pp. 123117–123133, 2021, doi: 10.1109/ACCESS.2021.3109894.
26. X. Song, K. Wang, L. Lei, L. Zhao, Y. Li and J. Wang, "Interference Minimization Resource Allocation for V2X Communication Underlaying 5G Cellular Networks", *Wireless Communications and Mobile Computing*, 2020, p. 19, Sep. 2020.
27. X. He, J. Lv, J. Zhao, X. Hou and T. Luo, "Design and Analysis of a Short-Term Sensing-Based Resource Selection Scheme for C-V2X Networks", *IEEE Internet of Things Journal*, 7(11), p. 1120911222, Nov. 2020.
28. A. Haider and S. H. Hwang, "Adaptive Transmit Power Control Algorithm for Sensing-Based Semi-Persistent Scheduling in C-V2X Mode 4 Communication", *Electronics, MDPI*, 8(8), p. 846, Jul. 2019.
29. R. Molina-Masegosa, M. Sepulcre and J. Gozalvez, "Geo-Based Scheduling for C-V2X Networks", *IEEE Transactions on Vehicular Technology*, 68(9), p. 83978407, Jun. 2019.
30. J. Kim, J. Lee, S. Moon and I. Hwang, "A Position-Based Resource Allocation Scheme for V2V Communication", *Wireless Personal Communications*, 98(1), p. 15691586, Jan. 2018.
31. F. Abbas, P. Fan and Z. Khan, "A Novel Low-Latency V2V Resource Allocation Scheme Based on Cellular V2X Communications", *IEEE Transactions on Intelligent Transportation Systems*, 20(6), p. 21852197, Jun. 2019.
32. W. Sun, D. Yuan, E. Ström and F. Brännström, "Cluster-Based Radio Resource Management for D2D-Supported Safety-Critical V2X Communications", *IEEE Transactions on Wireless Communications*, 15(4), p. 27562769, Apr. 2016.

8 Design of Smart Sensor-based Insole for Analysing Gait Parameters of Patients with Knee Arthroplasty

Sumit Raghav, Anshika Singh, Shashwat Pathak, Suresh Mani, Deepak Mittal, Mukul Kumar, and Anirudh Srivastava

8.1 INTRODUCTION

As a form of degenerative arthritis, Knee Osteoarthritis (KOA) is the most prevalent, resulting in the deterioration and degeneration of soft tissue and connective tissue [1]. Such peri-articular changes in the knee bring about a decrease in the motion and an aggravate stiffness and discomfort of the joint [2]. In the elderly population, knee osteoarthritis is a more common musculoskeletal condition. With a prevalence of 22% to 39% in India, it is the most common joint illness [3]. It is more common in women than in men, but with age the prevalence increases dramatically [4]. Almost 45% of women over 65 years of age have symptoms, while 70% of those over 65 years of age have radiological confirmation.

KOA, particularly among females, is a major cause of mobility impairment. In the elderly population, it also has a negative effect on overall well-being and quality of life [5]. Surgical procedures, pharmacological therapies, intra-articular injections and allied health services are among the treatments available for knee osteoarthritis. The most common surgical procedure performed to alleviate the pain and enhance function as well as overall experience of patients in the orthopaedic region is surgery, such as total knee arthroplasty. Norway, with Primary Knee Replacements of close to seven thousand in the year 2017 alone, reported an increased rate of total knee arthroplasty procedures [6].

Knee arthroplasty is supported by a rehabilitation programme of physiotherapy and exercises [7]. During the hospital stay, physical therapy concentrates on movement and independence from the bedside in a hospital. Moreover, therapeutic

DOI: 10.1201/9781003452645-8

assistance after surgery and individually planned exercise protocol support rebuilding and independence [8]. With an increase in life prospect, the anticipated rise in the prevalence of knee replacement, osteoarthritis, pre- and post-total knee replacement rehabilitation places a tremendous load on the healthcare system [9]. Regaining a normal gait pattern following knee replacement surgery is the key outcome of musculoskeletal rehabilitation [10].

Another approach for collecting and analysing results that are numerical and time-based is the laboratory-based gait assessment [11]. Identification of recurring walking issues has become easier due to the ability to analyse gait over time [12]. For patients without access to motion analysis labs, a device that can quantitatively assess gait might also be useful. Due to living in underdeveloped and rural areas with low income presents new potential for diagnosis and therapy for both physicians and patients. The entire process of measuring, quantifying and analysing a gait entails tracking a person's every step. It helps to determine the gait phase along with analysis of kinematics of related gait events for assessment of musculoskeletal functions.

Gait laboratories have reported motion capture systems through multiple cameras coupled with measurement of ground reaction forces [13]. The existing methods explained the need for a costly laboratory measurements and their available techniques [16]. Various motion-detection devices are available that are affordable, beyond the gait labs approach that incorporates the feasible and gait laboratory method's greatest features [14]. This chapter presents various breakthrough studies carried out on the creation and use of wearable sensor-based devices for gait analysis.

8.2   METHODOLOGY

A thorough search was conducted from 2015 to 2021 for obtaining inputs from existing technology to make the proposed design of a smart sensor-based insole. Many studies developed a novel sensor-centric technological programme to monitor gait parameters in clinical setups as well as home settings for clients with walking impairment for personalized feedback and also collected preliminary data of asymptomatic subjects. Many studies and clinical trials on the development and testing of wearable sensor-based technologies were released in the previous year. This chapter is based on a literature review covering the period between March 2015 and December 2021 with inclusion of papers that are subjective. These papers are discussed specifically, which are considered to offer new ideas and are of potential significance for clinical practice. An important option is to wait for new treatment modalities to appear to monitor joint replacements; data about the potential length of their last have become available [15].

Optimal treatment of knee osteoarthritis is still a particularly important subject of discussion. Gait analysis using external skin markers was used in the United States that provided a method to study kinematics and thus determine kinetic parameters on various total knee arthroplasty [16]. This method requires collection and proper analysis of collected data based on optimum algorithms. Furthermore, proper calibration is required to assure accurate results. For gait study, the Davis model is widely accepted and adopted for the placement of skin markers. Calibration requires static trials corresponding to specific body locations, to ensure quality measurement. So that the correct recording of walking pace and individual gait types is made, it is necessary to frequently check all acquired complicated datasets owing to exhaustion or attention. To obtain accurate analysis, these iterations should be repeated every three to five gait cycles, depending on the iteration. Relevant quantitative measurements are indulged with the information from the 3D marker collected from the focused analysis concerning adjudicate to the location of the joint centre and define the pinpoint of knee arthroplasty [17].

For the study of kinetic and kinematic variables shown in knee arthroplasty, this method provides better prospects for the aggregation and interconnection of precise, reliable, meticulous and consistent data [18]. Creating a model of the human gait in clinical practice is a lengthy and difficult procedure from an implementation standpoint. In order to overcome these difficulties, inertial sensors and adaptive algorithms are used here [19]. Analysing the readings of inertial sensors together with Artificial Intelligence algorithms have proven to be an excellent method to carry out a thorough gait analysis [20]. The majority of research projects found in the literature focused on asymptomatic clients. Therefore, more studies are required to enhance and further standardize the application in a variety of individuals. In order to create a suitable algorithm, scenario-specific study is a requirement [21].

Motion capture technology is frequently employed to measure human gait. The camera captures two-dimensional (2D) video because of its availability, and adaptability is the most popular method [22]. Yet, when compared to other ways, this method's effectiveness is lacking with standard gait analysis methods such as 3D gait analysis [23]. According to an analysis of the literature written between the years 1990 and 2019, about 30 research articles mention the relevant investigations. The complete requirements to do 2D gait monitoring in a relevant platform were described in this. Information on data-collecting procedures, data storage, specific age criteria, opted gait variables and other details are supplied with suggestions [22].

Following decades of development, measuring techniques for detailed gait analysis have become a useful approach for pinpointing specific gait problems. The high cost of operation along with the requirement of skilled manpower restricts the use of this approach to developed nations and financially secure groups of patients [24]. Therefore, observation-based study is still the most preferred method [25]. Recent advancements in low-cost wearable sensor technology, particularly inertial body sensors, cleared the path for a more effective replacement for observation-based methods. The fluid motion, flexibility of usage, accessibility and accuracy of the gait analysis made possible by these sensors immediately benefit the patient and the clinician by facilitating the collection of data and requiring less observation [26].

Human gait examination is a useful instrument to ensure prompt and precise detection of the disease, as well as a powerful tool for post-operation treatment and follow-up services. Capturing biomechanical parameters and relevant metrics in the environment of therapeutic, athletic and exploratory scenarios present another good method. These parameters are also categorized into musculoskeletal, neurological and circulatory, depending on the origin of the issue. These parameters are evaluated and discussed using different statistical/mathematical models. These studies further establish and strengthen the efficacy of gait analysis. Specifically, the measurement of parameters like a person's stride length, step width, gait speed, stance time and swing time plays a vital role in carrying out comprehensive gait analysis [27].

8.3 RESULTS

This study is designated to design a smart sensor-based insole by collecting the inputs from existing designs and developments that are implemented in healthcare setups. After reviewing the various available literature selected from the year 2015 to 2021 searched from different electronic databases, only 13 studies were included as shown in Table 8.1. The findings of the review are presented following the sequences such as author and year of publication, country and title of articles, parameters and findings. The need for early monitoring of the functions of the knee to substantiate the precise detection as soon as feasible and to immediately address the current condition of the issue is stressed in previous publications.

There are various designs executed to develop the wearable sensor insole device for analysing the walking problems of patients with systemic disorders. In order to evaluate human mobility and compile quantitative evidence for real-time data, gait analysis is a crucial step. However, to set up a gait lab, a sizable space is needed, along with pricey equipment. A large sum of money is required to construct and establish a gait lab [28]. Before and after knee therapy, patients who had trouble walking are increasingly being evaluated and helped by portable technologies and remote-centric methods in an effort to lessen the cost load [29].

8.4 DISCUSSION

However, among the most existing designs of sensor base insole that are capable enough to capture gait events in and out the clinical settings [30]. Despite the fact that instrumented gait labs are reliable or trustworthy to monitor gait patterns and approach is steady for longer time until the user might change the shoe during analysis of gait because these highly instrumented labs may record the user's interactions with their surroundings [31].

Technologies such as force sensor mats or walkways, smart shoes with embedded sensors may be used to map plantar pressure and measure gait characteristics [32]. However, these technologies are limited in some ways to a particular field, making it impractical for every patient to track their daily motor activities and associated metrics. Therefore, we are convinced that portable sensor-induced technology should be used instead of instrumented gait labs as the primary measurement tool to identify and assess individuals with abnormal gait metrics after knee arthroplasty [33].

To provide such machine learning-driven devices to be used by individuals in everyday life is an utmost need and a major issue. Individuals with mobility impairment have some challenges with rehabilitation and technology-assisted devices with respect to their uses in the dynamic status of the lower limb. In order to assess functional status during and following rehabilitation for various pathologies, some clinical parameters have been recommended [34].

TABLE 8.1

Summary of Included Studies that Categorized in Author and Year of Publication, Country, Title of the Study, Parameters Used and Main Findings of the Study

Sr. No.	Author and Year of Publication	Country	Title	Parameters Used	Main Findings
1.	Y. S. Ashad Mustufa et. al (2015)	Cambridge (USA)	Design of a Smart Insole for Ambulatory Gait Analysis	Gait Variables, Foot Mapping	Several Integrated Sensors Showcase a Crucial Function in Detecting Plantar Pressure and Gait Phases
2.	Benedikt Johannes Braun et al. (2015)	Germany	Validation and Reliability Testing of a New, Fully Integrated Gait Analysis Insole	Kinetics, Parameters of gait	A Brand-New Tool (Opengo, developed by Moticon Gmbh) was released to continuously detect data on the kinetics and spatiotemporal characteristics of gait. It can be used in clinical trials for an extended time interval.
3.	Adin MingTan et al. (2015)	Australia	Design of Low-Cost Smart Insole for Real-Time Measurement of Plantar Pressure	Foot mapping	This innovative insole offers real-time presentation of the sole's pressure mapping and considered for therapy and athletic activity evaluation.
4.	Jeeshan Rahman, et al. (2015)	London (UK)	Gait Assessment as a Functional Outcome Measure in Total Knee Arthroplasty: A Cross-Sectional Study	Parameters of gait, Knee angles	Inertial measurement units were used to measure the knee angle, limb segment angle and gait parameters. Less evidence has been found about the improvement of gait variables 12 months after knee replacement.
5.	Patsiri Wannaphan and Teeranoot Chanthasope- phan (2016)	Thailand	Position Controlled Knee Rehabilitation Orthotic Device for Patients after Total Knee Arthroplasty	Arc of knee joint	Patients who have undergone knee arthroplasty show a significant improvement in their range of motion and their walking speed thanks to a torque feedback controlled device.
6.	Payal S. Malvade; Atul K. Joshi; Swati P. Madhe (2017)	India	In-sole Shoe Foot Pressure Monitoring for Gait Analysis	Gait Parameters, Plantar pressure	Real-time plantar pressure and movement are detected by an insole with force-sensitive resistor embedded technology. It could be beneficial for clinical gait analysis

(Continued)

TABLE 8.1 (CONTINUED)

Summary of Included Studies that Categorized in Author and Year of Publication, Country, Title of the Study, Parameters Used and Main Findings of the Study

Sr. No.	Author and Year of Publication	Country	Title	Parameters Used	Main Findings
7.	Hongyu Zhao; Zhelong Wang; Sen Qiu; Yanming Shen; Jianjun Wang (2017)	China	IMU-Based Gait Analysis for Rehabilitation Assessment of Patients with Gait Disorders	Gait variables	When examining patients with gait disorders, dual foot-mounted inertial measurement units were calibrated to pick up on the gait variables. During their clinical practice, healthcare professionals might also find it helpful.
8.	Katharina Gordt, Thomas Gerhardy, Bijan Najafi, Michael Schwenk (2018)	Germany	Effects of Wearable Sensor-Based Balance and Gait Training on Balance, Gait, and Functional Performance in Healthy and Patient Populations	Gait variables and balance	There is some evidence that some gait parameters and balance measurements can be trained with wearable sensors.
9.	Nils Roth et al (2018)	Germany	Synchronized Sensor Insoles for Clinical Gait Analysis in Home-Monitoring Applications	Gait parameters and foot mapping	Utilizing the GaitRite system as a reference, the gait parameters and plantar pressure are measured using a low power sensor insole. It demonstrated its capacity to acquire coordinated gait data and address the demand for monitoring at home over clinical setup.
10.	Armelle M. Ngueleu et al. (2019)	Canada	Design and Accuracy of an Instrumented Insole using Pressure Sensors for Step Count	Cadence, Plantar pressure	The development and use of the force sensitive resistor Integrate Insole allowed for the real-time detection of step count. The Cumulative Sum-Based Method Displays a High Degree of Accuracy.
11.	Armelle M. Ngueleu, et al. (2019)	Canada	Validity of Instrumented Insoles for Step Counting, Posture and Activity Recognition	Gait variables, Postural analysis in activities	For step counting, the instrumented insole appeared surprisingly steady, but was not capable/reliable to capture posture and body position.

(Continued)

TABLE 8.1 (CONTINUED)

Summary of Included Studies that Categorized in Author and Year of Publication, Country, Title of the Study, Parameters Used and Main Findings of the Study

Sr. No.	Author and Year of Publication	Country	Title	Parameters Used	Main Findings
12.	Xia Wang, David J et al. (2019)	Australia	Technology-Assisted Rehabilitation Following Total Knee or Hip Replacement for People with Osteoarthritis	Pain scales, Gait variables	It demonstrates how technology-driven therapy significantly reduced pain while having less of an impact on function.
13.	Hari Prasanth, et al. (2021)	Netherlands	Wearable Sensor-Based Real-Time Gait Detection	Gait events	This study suggests that combining IMU and rule-based methods is the best approach for real-time data of gait analysis.

When taking into account the widespread use of sensor-based portable devices, such as smart shoes, smart watches and smart insoles, they offer positive and potentially positive results in support of the validity and reliability of data gathered and analysed during assessment and therapy [35]. However, all of these portable wearable devices have some limitations when it comes to fully monitoring all parameters of gait. There are just a few clinical fields that offer this type of technology-assisted rehabilitation, making it unaffordable for everyone who experiences problems with their motion and muscle recruitment following joint surgery in the lower extremities [36]. For patients who belong to remote areas, technology-assisted rehabilitation is especially required because they confront a variety of financial difficulties and may choose not to visit the hospital frequently to have their health conditions assessed [37]. So, there is a need to develop and establish tele-rehabilitation service featured with technology-assisted rehabilitation that can be helpful to empower patients in their walking patterns [38].

In the current environment, sensor-based technology that can assist physicians and patients while also playing a significant role in the healthcare system is needed. This technology is valid and seems more reliable than standard devices like instrumented gait labs. These technologically based interventional approaches have the potential to replace established methods in the future for both practitioners and patients in both clinical and home settings. However, given that this type of technology-based assessment is still costly and challenging for everyone to use and maintain, it is necessary to focus on its cost and feasibility [39]. A big population can profit from this technology if it is built with a low budget and simpler management requirements, and it could set a standard for the healthcare field.

8.4.1 Proposed Design

Currently, evaluating gait requires a well-equipped expensive and complex gait biomechanics lab. To date, evaluation based on ground reaction forces with the force plate is the gold standard for detecting the initial contact and toe off. In the last decade, there was a high demand for wearable sensors which are applications based on smartphones for gait analysis outside laboratory settings in the real world. The patients of knee arthroplasty show considerable changes in the gait parameters. It becomes an important criterion to test the validity and reliability of Smart Sensor-Based Insole (SSBI) approaches for detecting abnormalities in the gait parameters. SSBI will have the following major components as in Table 8.2.

The first component is a rubber-made insole of the prototype of SSBI embedded with low-cost sensors for evaluation of gait parameters including force-sensitive resistors as pressure sensors and Microprocessor Unit (MPU) (accelerometer, gyroscope and magnetometer) as motion sensors. The pressure sensors will be used to obtain the pressure map under the foot. MPU embedded with accelerometer, gyroscope and magnetometer will give the information regarding the movements of the foot as in Figure 8.1 a and 8.1 b. The second stage of the proposed tool will be the module for acquisition and transmission of the signals.

TABLE 8.2

The Components will Be Used to Develop Smart Sensor-Based Insole

Type of Sensor	Parameters	Output of Sensor
Accelerometer	Stride length, Stride velocity and displacement	Voltage change corresponding to acceleration
Gyroscope	Orientation	Voltage change corresponding to angular velocity
Magnetometer	Navigation	Dynamic measurement range and high resolution with lower current consumption
Pressure sensor (FSR)	Pressure distribution under the foot, heel – strike timing and heel-off timing	Resistance change corresponding to applied pressure across the sensor, resulting from change in compression of the sensor

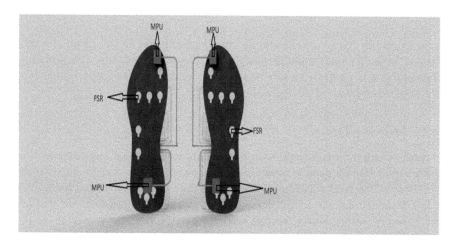

FIGURE 8.1A Computer-Aided Design (CAD) of smart sensor-based insole (front-side)

Bluetooth will be used to transmit the data from insole to the smartphone. A chargeable battery (9 volt) will be fitted into the insole to provide the electrical supply. The third stage of the proposed tool will be the module for aggregation and processing of the collected data of gait. A smartphone application will be deployed to compare the collected data of gait with the baseline data. The smartphone application will give audio-visual feedback to the healthcare provider as well as to the patient regarding the gait parameters. Various studies have been done on the development and use of wearable devices for analysing the human gait. However, none of them focused on inventing a device which can provide personalized coaching to the patients in terms of evaluating and correcting the altered gait parameters.

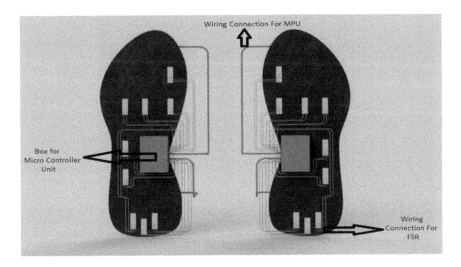

FIGURE 8.1B   Computer-Aided Design (CAD) of smart sensor-based insole (back side)

This research will focus on designing and developing a SSBI for patients with knee arthroplasty.

8.5 CONCLUSION

This study concludes that there is a high amount of work done so far in terms of the development of wearable devices named intelligent sole, orthotic insole, smart

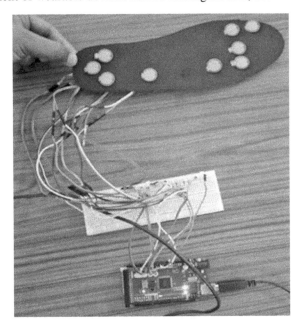

FIGURE 8.2A   Prototype of smart sensor-based insole

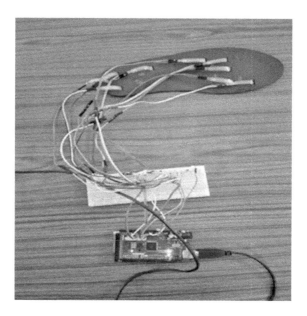

FIGURE 8.2B   Prototype of smart sensor-based insole

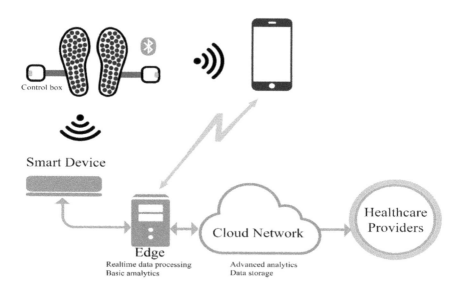

FIGURE 8.3   Proposed methodology for analysing and collecting the data of gait [38]

insole, smart shoe etc. for analysing gait. These devices are available in developed countries and are widely used. Due to lack of awareness this technology is not recognized in the healthcare system in India. There is high demand to develop these portable devices and to provide them to users at affordable prices in the market in India. Considering all challenges and opportunities, we have decided to design a

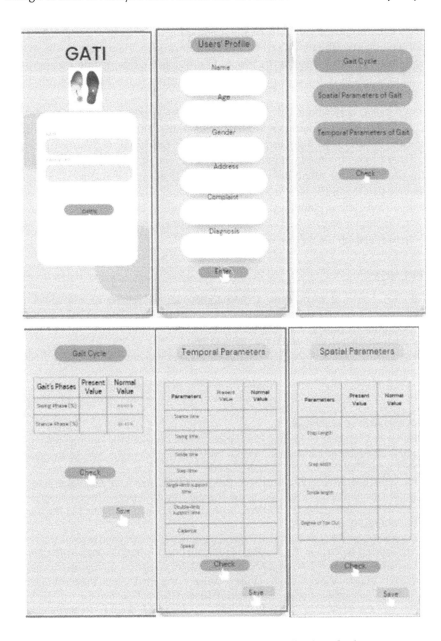

FIGURE 8.4   Web-based application display for data visualization of gait

SSBI for improving the gait parameters in patients with knee arthroplasty. We have also developed a prototype of SSBI considering the proposed design (Figure 8.2 a and 8.2 b). In this prototype, force-sensitive resistors for plantar pressure mapping, and 9-axis Accelerometer, gyroscope and magnetometer for monitoring the gait variables will be calibrated in the proposed design of the SSBI. The second process of the proposed tool will be the module for the acquisition and transmission of the signals. Bluetooth will be used to transmit the data from insole to the smartphone. The third stage of the proposed tool will be the module for aggregation and processing of the data of gait and transmitting that data to a smartphone as in Figure 8.3.

A web-based application will be used to show and compare the collected data with the baseline/user's data (Figure 8.4). This application will give audio-visual feedback to the user regarding the gait parameters. In the initial stage, we will collect data of gait of asymptomatic individuals with different anthropometric characteristics to substantiate the reliability and validity of this device for further application. We assume that this SSBI will be a good approach for patients with mobility impairment, due to different pathology, to get early mobility. Initial monitoring via the proposed SSBI can be executed to achieve optimum conclusion and expected result and help to speed-up the recovery rate.

REFERENCES

1. Ku YH, Lee H, Ryu HY, et al. A clinical pilot study to evaluate the efficacy of oral intake of Phellinus linteus (sanghuang) extract on knee joint and articular cartilage: Study protocol clinical trial (SPIRIT Compliant). *Med (United States)*. Epub ahead of print 2020. DOI: 10.1097/MD.0000000000018912.
2. Meyer AM, Thomas-Aitken HD, Brouillette MJ, Westermann RW, Goetz JE. Isolated changes in femoral version do not alter intra-articular contact mechanics in cadaveric hips. *J Biomech*. Epub ahead of print 2020. DOI: 10.1016/j.jbiomech.2020.109891.
3. Gupta G. Prevalence of musculoskeletal disorders in farmers of Kanpur-Rural, India. *J Community Med Health Educ*. Epub ahead of print 2013. DOI: 10.4172/2161-0711.1000249.
4. Jadhav S, Dudhekar U, Saoji K, Chaudhari SS. Outcome analysis of high tibial osteotomy in osteoarthritis of knee: A study protocol. *Int J Curr Res Rev*. Epub ahead of print 2020. DOI: 10.31782/IJCRR.2020.SP60.
5. Bindawas SM, Vennu V, Alfhadel S, Al-Otaibi AD, Binnasser AS. Knee pain and health-related quality of life among older patients with different knee osteoarthritis severity in Saudi Arabia. *PLOS ONE*. Epub ahead of print 2018. DOI: 10.1371/journal.pone.0196150.
6. Kurtz SM, Ong KL, Lau E, et al. International survey of primary and revision total knee replacement. *Int Orthop*. Epub ahead of print 2011. DOI: 10.1007/s00264-011-1235-5.
7. Zapparoli L, Sacheli LM, Seghezzi S, et al. Motor imagery training speeds up gait recovery and decreases the risk of falls in patients submitted to total knee arthroplasty. *Sci Rep*. Epub ahead of print 2020. DOI: 10.1038/s41598-020-65820-5.
8. Zeni J, Logerstedt D, Flowers P, Abujaber S, Snyder-Mackler L. Rehabilitation to reduce secondary osteoarthritis after total knee arthroplasty. *Osteoarthr Cartil*. Epub ahead of print 2012. DOI: 10.1016/j.joca.2012.02.444.
9. Nemes S, Rolfson O, W-Dahl A, et al. Historical view and future demand for knee arthroplasty in Sweden. *Acta Orthop*. Epub ahead of print 2015. DOI: 10.3109/17453674.2015.1034608.
10. Mohapatra S, Cheung KL, Hiligsmann M, Anokye N. Most important factors for deciding rehabilitation provision for severe stroke survivors post hospital discharge: A study protocol for a best–worst scaling experiment. *Methods Protoc*. Epub ahead of print 2021. DOI: 10.3390/mps4020027.

11. Marýn J, Blanco T, Marín JJ, et al. Integrating a gait analysis test in hospital rehabilitation: A service design approach. *PLOS ONE*. Epub ahead of print 2019. DOI: 10.1371/journal.pone.0224409.

12. Papi E, Bo YN, McGregor AH. A flexible wearable sensor for knee flexion assessment during gait. *Gait Posture*. Epub ahead of print 2018. DOI: 10.1016/j.gaitpost.2018.04.015.

13. Wang C, Chan PPK, Lam BMF, et al. Real-time estimation of knee adduction moment for gait retraining in patients with knee osteoarthritis. *IEEE Trans Neural Syst Rehabil Eng*. Epub ahead of print 2020. DOI: 10.1109/TNSRE.2020.2978537.

14. Lee JK, Han SJ, Kim K, Kim YH, Lee S. Wireless epidermal six-axis inertial measurement units for real-time joint angle estimation. *Appl Sci*. Epub ahead of print 2020. DOI: 10.3390/app10072240.

15. Wallis JA, Taylor NF. Pre-operative interventions (non-surgical and non-pharmacological) for patients with hip or knee osteoarthritis awaiting joint replacement surgery - A systematic review and meta-analysis. *Osteoarthr Cartil*. Epub ahead of print 2011. DOI: 10.1016/j.joca.2011.09.001.

16. Fujimaki Y, Miyawaki M, Thorhauer E, et al. In vivo kinematics of the ankle during gait following reconstruction for chronic ankle instability. *Arthrosc J Arthrosc Relat Surg*. Epub ahead of print 2013. DOI: 10.1016/j.arthro.2013.07.054.

17. Jain R, Kalia RB, Das L. Anthropometric measurements of patella and its clinical implications. *Eur J Orthop Surg Traumatol*. Epub ahead of print 2019. DOI: 10.1007/s00590-019-02490-8.

18. Papagiannis GI, Triantafyllou AI, Roumpelakis IM, Papagelopoulos PJ, Babis GC. Gait analysis methodology for the measurement of biomechanical parameters in total knee arthroplasties: A literature review. *J Orthop*. Epub ahead of print 2018. DOI: 10.1016/j.jor.2018.01.048.

19. Ahmed A, Roumeliotis S. A visual-inertial approach to human gait estimation. In *Proceedings - IEEE International Conference on Robotics and Automation*. 2018. Epub ahead of print 2018. DOI: 10.1109/ICRA.2018.8460871.

20. Zou Q, Ni L, Wang Q, Li Q, Wang S. Robust gait recognition by integrating inertial and RGBD sensors. *IEEE Trans Cybern*. Epub ahead of print 2017. DOI: 10.1109/TCYB.2017.2682280.

21. Alkhatib R, DIab MO, Corbier C, Badaoui ME. Machine learning algorithm for gait analysis and classification on early detection of Parkinson. *IEEE Sens Lett*. Epub ahead of print 2020. DOI: 10.1109/LSENS.2020.2994938.

22. Michelini A, Eshraghi A, Andrysek J. Two-dimensional video gait analysis: A systematic review of reliability, validity, and best practice considerations. *Prosthet Orthot Int*. Epub ahead of print 2020. DOI: 10.1177/0309364620921290.

23. Teufl W, Taetz B, Miezal M, et al. Towards an inertial sensor-based wearable feedback system for patients after total hip arthroplasty: Validity and applicability for gait classification with gait kinematics-based features. *Sensors (Switzerland)*. Epub ahead of print 2019. DOI: 10.3390/s19225006.

24. Mukaino M, Ohtsuka K, Tanikawa H, et al. Clinical-oriented three-dimensional gait analysis method for evaluating gait disorder. *J Vis Exp*. Epub ahead of print 2018. DOI: 10.3791/57063.

25. Brunnekreef JJ, Van Uden CJT, Van Moorsel S, Kooloos JG. Reliability of videotaped observational gait analysis in patients with orthopedic impairments. *BMC Musculoskelet Disord*. Epub ahead of print 2005. DOI: 10.1186/1471-2474-6-17.

26. Pierleoni P, Pinti F, Belli A, et al. A dataset for wearable sensors validation in gait analysis. *Data Br*. Epub ahead of print 2020. DOI: 10.1016/j.dib.2020.105918.

27. Sansgiri S, Visscher R, Singh NB, et al. A comparison of clinically and kinematically identified spatio-temporal parameters in cerebral palsy gait. *Gait Posture*. Epub ahead of print 2020. DOI: 10.1016/j.gaitpost.2020.08.057.

28. Seichert N, Senn E. Clinical meaning of the torque between stance leg and ground for the analysis of gait mechanism. *Clin Investig*. Epub ahead of print 1993. DOI: 10.1007/BF00180104.

29. Kim I, Heo JS, Hossain MF. Challenges in design and fabrication of flexible/stretchable carbon-and textile-based wearable sensors for health monitoring: A critical review. *Sensors (Switzerland)*. Epub ahead of print 2020. DOI: 10.3390/s20143927.

30. Nagano H, Begg RK. Shoe-insole technology for injury prevention in walking. *Sensors (Switzerland)*. Epub ahead of print 2018. DOI: 10.3390/s18051468.

31. Marsan T, Rouch P, Thoreux P, Jacquet-Yquel R, Sauret C. Estimating the GRF under one foot knowing the other one during table tennis strokes: A preliminary study. *Comput Methods Biomech Biomed Eng*. Epub ahead of print 2020. DOI: 10.1080/10255842.2020.1813422.

32. Virmani T, Gupta H, Shah J, Larson-Prior L. Objective measures of gait and balance in healthy non-falling adults as a function of age. *Gait Posture*. Epub ahead of print 2018. DOI: 10.1016/j.gaitpost.2018.07.167.

33. Renner K, Queen R. Detection of age and gender differences in walking using mobile wearable sensors. *Gait Posture*. Epub ahead of print 2021. DOI: 10.1016/j.gaitpost.2021.04.017.

34. Chang KW, Lin CM, Yen CW, et al. The effect of walking backward on a treadmill on balance, speed of walking and cardiopulmonary fitness for patients with chronic stroke: A pilot study. *Int J Environ Res Public Health*. Epub ahead of print 2021. DOI: 10.3390/ijerph18052376.

35. Drăgulinescu A, Drăgulinescu AM, Zincă G, et al. Smart socks and in-shoe systems: State-of-the-art for two popular technologies for foot motion analysis, sports, and medical applications. *Sensors (Switzerland)*. Epub ahead of print 2020. DOI: 10.3390/s20154316.

36. Chaurasia DID, Shukla DS, Gupta DA, et al. Outcome of the unidentified/unaccompanied patient of traumatic brain injury in trauma unit of Gandhi Medical College and associated Hamidia hospital Bhopal (India). *Int J Med Biomed Stud*. Epub ahead of print 2019. DOI: 10.32553/ijmbs.v3i11.749.

37. Yoo DH, Kim SY. Effects of upper limb robot-assisted therapy in the rehabilitation of stroke patients. *J Phys Ther Sci*. Epub ahead of print 2015. DOI: 10.1589/jpts.27.677.

38. Plaza A, Fabà M, Inzitari M, et al. The Return Home Program: Integrated health and social care for post-stroke patients. *Int J Integr Care*. Epub ahead of print 2016. DOI: 10.5334/ijic.3029.

39. Rodrigues TB, Salgado DP, Catháin C, O'Connor N, Murray N. Human gait assessment using a 3D marker-less multimodal motion capture system. *Multimed Tools Appl*. Epub ahead of print 2020. DOI: 10.1007/s11042-019-08275-9.

9 Evaluation of Female Facial Attractiveness and Smile Aesthetics Using Eye-Tracking System
A Photographic Study

Swabhiman Behera, Kamlesh Singh, Shashwat Pathak, Ragni Tandon, and Pratik Chandra

INTRODUCTION

In the past years, great emphasis has been given to facial attractiveness for both social and professional interactions, where smile aesthetics plays an important role. Facial attractiveness is affected by various components which include eyes, mouth, ears, nose, hairstyle and complexion. In most professional fields aesthetics of a person plays a huge role in determining their self-confidence and their interaction with people around them. So, now most patients seek orthodontic treatment not only for improving their mal-aligned teeth and other dental problems but also to improve the overall facial and smile aesthetics [1, 2]. Recent studies have shown that the perception of facial attractiveness and smile aesthetics have developed and there is a significant change among laypersons and professionals. Different people and even the sex of an individual effects how they perceive a face when they look at a person [3–5]. There are many studies on aesthetics and smile where the subjective assessment of a smile is perceived visually by examining constructed images with respect to the visual analogue scale. These studies mainly contained questionnaire which were circulated among peers to obtain suitable data. But this technique has been going on since the 1900s. Today in the era of Artificial Intelligence (AI), it is necessary to utilize the resources and obtain objective data regarding which area of the face and smile is most emphasized [6, 7].

The recent advancement in technology has contributed to the field of research and studies, an eye-tracking system with the help of which we can collect objective data featuring which location viewers view most frequently (fixation density), and how much time they spend viewing those areas (fixation duration). An eye tracker is

DOI: 10.1201/9781003452645-9

a device that measures and detects the movement of the eyes by tracking the corneal reflection. This technology is mainly used in the field of research, psychology, in marketing and human computer interaction on different platforms [8].

Eye-tracking system is an objective method for detecting where a person's visual attention is concentrated. The system consists of a camera and video-processing software to obtain the quantitative measure of a person's real-time visual attention. The pupil is tracked using infrared or near-infrared light, and corneal reflection is utilized for visual attention recording [9, 10].

A recent study by Richards et al. conducted on female facial attractiveness through eye tracking showed significant differences in the way people perceive attractiveness in female subjects. They concluded that a female with an unattractive smile grabbed more attention compared to other facial areas in an average and attractive face [11].

The reliability of eye-tracking system has been witnessed in a few other studies too [12–14]. Hence the purpose of this study is to evaluate the difference in subjective and objective perception of smile aesthetics and overall facial attractiveness, respectively, among various professional fields through an eye-tracking system and to correlate between visual and verbal expression of ROI. In future, this software can play a vital role in "Smile Designing" which is a trending and desirable treatment plan for most of the patients seeking treatment only for their smile and aesthetics. Furthermore, this software could also be used to boost the psychological and self-confidence of a person.

After successful intervention of all the components of the software, it can be commercialized to gain revenue and will generate a strong base for researchers from the field of dentistry as well as psychology.

METHODOLOGY

PARTICIPANTS

A total of 200 female participants were recruited for clicking photographs who satisfied our inclusion criteria and were willing to participate in the study.

- Young adults (age 18–30 years).
- No facial asymmetry.
- No previous orthodontic treatment.
- North Indian population.
- Not wearing any makeup or willing to remove it.
- No piercing.
- No facial tattoo.

The photos were further evaluated by experienced orthodontists to remove any imaging error such as blurred image, subject not looking directly into the lens, or untied hairs affecting the face; a total of 50 photos were selected. Out of the 50 photos, 10 photos were randomly selected for inclusion in the study.

Posed smile photographs of all the female participants were clicked with a professional mirrorless camera (Sony A6400, Tokyo, Japan with Sigma 56mm f/1.4 DC DN Contemporary Lens, Fukushima, Japan) that was mounted on a tripod and assisted by artificial flash light which were set on a fixed intensity. Adobe Photoshop CC 2019 software (Adobe Systems Inc, San Jose, Calif) was used to crop and obtain the posed smile image and get two photographs, i.e., full face with smile and dental smile. These photos were standardized using the same software.

A total of 20 photos were obtained (10 full face, 10 dental smile) of the selected 10 subjects. These photos were arranged randomly in a presentation slideshow. A transition slide was placed between each image to reset the eyes of the viewers. A timer was set on each image slide of 6 seconds and 3 seconds for the transition slide.

A total of 6 classes of people were selected and each class had 10 participants for the survey. The inclusion criteria for these participants were, they should understand English, should have no prior neurological condition, should be aged between 18 and 35 years, should not wear any kind of lenses, should not have colour blindness and no recent use of alcohol or any other drug that would affect cognitive abilities.

Each subject was seated 60–70 cm away from a 17-inch screen, the camera for eye tracking was kept below the screen. Four practice images were shown before starting the real study so that the participants get used to the manner of image they will be shown. Regions of Interest (ROI) of the face defined where the viewer's gaze paused for 80 ms or longer (a fixation) when viewing each image, in essence creating a map of the face. These areas were forehead, hair, eyebrows, eyes, nose, mouth, cheeks, chin and ears. A small gap was left between interest areas to ensure accuracy of a fixation. After the eye-tracking session the participants were shown the same slide show but this time they were asked to rate the images in a VAS.

LITERATURE REVIEW

Robin S. Baker et al. in 2018 [9] conducted a study using an eye-tracking system in a similar manner as done by Richards et al. and Johnson et al. But he examined only male models to find out whether there is a point on the AC-IOTN grade where dental attractiveness would directly affect the background facial attractiveness. He constructed a total of nine composite images which were edited and combined with different levels of AC-IOTN determined previously. These images were viewed by 64 raters through eye tracking and data analysis was done. The viewers mainly concentrated on the mouth of an average attractive man. In IOTN level 7 the eye and mouth gained similar attention, as the IOTN level increased to 10 the unesthetic malocclusion attracted the viewers' attention. Overall, for a lay person both male and female their attention shifts towards the mouth in case of moderate to severe malocclusion and this is independent of background attractiveness.

Michael R. Richards et al. in 2015 [10] conducted a study to evaluate whether different levels of dental and facial attractiveness changes visual attention in term of the area of interest, the area viewed first, the greatest number of times and the longest time by using the eye-tracking system. They selected 24 composite images of female subjects who were graded into three grades, i.e., 1, 7, 10 according to AC-IOTN. These images were a combination of dental and facial attractiveness. The images were shown to 76 viewers through eye tracking to know the frequency, time and location while viewing the faces. From the data collected, they found out that eye tracking is a reliable source for objective evaluation, eyes were more important than mouth in terms of both duration and fixation, grade of dental attractiveness affected the way viewers looked at faces and background facial attractiveness was an independent factor and sex of the viewers affected the study.

Wang et al. in 2016 [11] used eye tracking to evaluate the perspective of a layperson among normal occlusion and malocclusion and to record the scan path and assessment of pre-treatment and post-treatment patient photographs. They selected photographs of 20 patients from their archives whose treatment was completed. These photographs were judged by 88 participants (laypeople). The results showed that in patients with malocclusion the manner of gaze is different. Post-orthodontic treatment normalizes the scan path similar to a normal person with good occlusion and no orthodontic treatment.

Elizabeth K. Johnson et al. in 2017 [12] conducted a study using eye-tracking system based on AC-IOTN ranging from grade 3–7. They selected 15 composite images by combining 3 facial attractiveness levels and 5 levels of dental attractiveness. These images were subjected to eye tracking by 66 lay participants. Laypersons concentrated more on the eyes and as AC-IOTN decreased their attention shifted towards the mouth. They concluded that as dental attractiveness decreases the attention shifts more towards the mouth from the eyes and other features.

Murat Celikdelen et al. in 2020 [12] conducted a study to evaluate the effect of buccal corridor width, gingival display and upper midline deviation on smile aesthetics. The photos of 15 smiling female participants were taken. Each photo was evaluated by 8 orthodontists to select 1 photo that represented the general facial features of our society. The selected photo was manipulated in a photo-editing software to achieve 21 photos. Each photo was viewed by 16 laypeople who underwent the eye-tracking study. From the eye-tracking data, it was found that the difference in the buccal corridor did not influence the attractiveness score and the gaze duration. But a gingival display of 2 mm and above gained attention in females and 5 mm and above was observed more in males. The change in the midline was not significant for laypersons even at a 6-mm deviation.

de Oliveria et al. in 2019 [13] conducted a study to determine whether different classes of observers view occlusal changes differently, to determine whether all groups rate similarly, and how different grades of malocclusion change the viewers' attention. They included three facial images which were segregated into three groups of IOTN by experienced orthodontists. Each image was viewed for 59 seconds and a VAS scale was put after each image to obtain the subjective data. Ninety participants were distributed into three groups – dentists, orthodontists and laypeople viewed these images through eye tracking. From the data collected it was concluded that there was a significant difference in how the images were evaluated by the different groups, for grade 8 IOTN orthodontists analysed the dental problem more closely than dentists and laypersons.

Kim et al. in 2018 [15] in their research to find out how the viewer's perception of facial attractiveness changes according to various facial angle and smile recruited a total of 33 young adults and 6 models (3 men, 3 women) to provide their facial photos. They arranged different photos in a single panel and each participant was asked to evaluate them through eye tracking as well and on-screen scoring. They included 3 tasks: Task 1 – rating of the overall facial attractiveness; Task 2 – select the most attractive face; Task 3 – evaluate how facial angle and smile changes the rating of facial attractiveness. After evaluation of the results, it was found that among neutral model faces the 0° face showed the maximum fixation time. The most attractive looking face was at 30° and 45° for both model and self faces 0° face had the highest fixation time.

Peishan Huang et al. in 2019 [16] studied the attention of laypeople on different facial profiles with altered mandibular protrusion and the effect of facial background attractiveness. Profile photos of 200 subjects were clicked and standardized using editing software. Forty volunteers were recruited to rate the attractiveness of the photographs and they were arranged according to the mean attractiveness score. Finally, a total of 24 different images were selected. Fifty-five participants were recruited and the images were viewed to them through eye-tracking system. From this study it was found that for a normal profile or slightly protruded profile eyes were the most viewed region, as the mandibular protrusion increased the AOI shifted from eye, nose, cheeks towards the lower face. The visual attention of the viewers was affected by the mandibular protrusion as well as background facial attractiveness.

Trevisan et al. in 2020 [17] studied the visual perception of 60 participants: 30 dentistry students (15 male, 15 female) and 30 laypeople (15 male, 15 female) for different levels of maxillary central incisor abrasion. They included six altered images of people with different grades of abrasion. All the 60 participants underwent the eye-tracking session and it was noted that for the dental students the greatest amount of focus was towards the central incisors and slightly diverged towards the eye, whereas for the layperson the main focus was diverted towards the eyes even when their attention was on the abrasion.

Predrag N. et al. in 2020 [18] recorded the subjective data of 15 orthodontists, 15 final-year dental students and 15 laypeople on 21 smile photographs of White people to understand the difference in the pattern of evaluation by all the three categories. He concluded that according to the orthodontist, crowding and asymmetry affect the aesthetic more than any other feature. There was no significant difference in the evaluation by both dental students and laypeople. It is important to make proper protocols in diagnosis and treatment planning for a patient as there is a large difference in the viewing pattern of the orthodontist and a layperson.

Rafael B. Wolanski et al. [19] in 2020 studied the perception of mandibular laterognatism and smile aesthetics using the eye-tracking system. Photograph of a male and a female subject was taken in frontal view while smiling and it was standardized using editing software. Four altered images of each subject were created. Sixty laypeople (30 men and 30 women) were considered as raters who underwent the eye-tracking session for the eight images, their eye movements were recorded and it was found that untrained laypeople can recognize laterognatism and understand facial aesthetics. Females with laterognatism towards the right gained more attention than malse with laterognatism towards the left.

From the above literature we can conclude that the best way to understand the psychological aspect of a person is to undergo eye tracking, rather than a verbal or a questionnaire survey. Moreover a reliable and accurate method could be to use both questionnaire as well as eye tracking to understand both subjective as well as objective intentions of a viewer.

Table 9.1 gives details for the above statement.

RESULTS

Visual exploration of faces help to understand how individuals view various faces. There is a chance of deception in a questionnaire survey, so with the help of eye tracking we will reveal the real-time data of a person's gaze (Figures 9.1 and 9.2).

Statistical analysis required = ANOVA and post hoc testing with Tukey-Kramer procedure.

Data required from eye tracking = f {Fixation Duration, Fixation Density (Number of fixation)}

TABLE 9.1
List of Major Works and their Findings

Sl No.	Researcher	Year	Method	Accuracy
1	Michael R. Richards et al.	2015	Eye tracking (EyeLink 1000; SR Research)	Good
2	Wang et al.	2016	Eye tracking (EyeLink 1000; SR Research)	Good
3	Elizabeth K. Johnson et al.	2017	Eye tracking (EyeLink 1000; SR Research) questionnaire	Excellent
4	Robin S. Baker et al.	2018	Eye tracking (EyeLink 1000; SR Research)	Good
5	Seol Hee et al.	2018	SMI RED 120 eye-tracker (SensoMotoric Instruments) Questionnaire	Excellent
6	de Oliveria et al.	2019	The Eye Tribe Tracker sensor (Copenhagen, Denmark)	Good
7	Trevisan et al.	2020	The Eye Tribe Tracker sensor (Copenhagen, Denmark)	Good
8	Rafael B. Wolanski et al.	2020	The Eye Tribe Tracker sensor (Copenhagen, Denmark)	Good

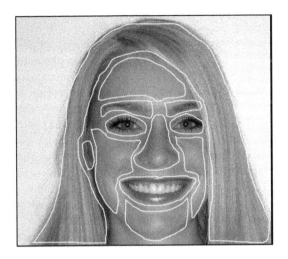

FIGURE 9.1   Region of interest in face (ROI): mouth, eyes, nose, chin, hair

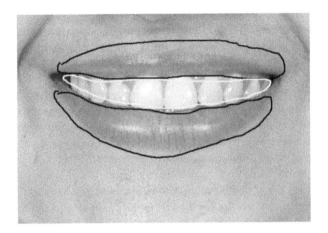

FIGURE 9.2   Region of interest in smile (ROI – smile): lips, teeth, smile curve

Questionnaire for the viewers after eye tracking for each image they viewed (20 images: 10 full face, 10 smile) included the following two points:

(i) Score of attractiveness out of 10?
(ii) Which part did you view while judging?

Experiments were performed on Python 3.11.1 in Windows and mentioned features (methodology section) were observed and extracted for further analysis. Results obtained were plotted using Python (Jupyter) and are presented in this section for detailed analysis.

Average of total 10 subjects for the various codes

DISCUSSION

In today's era, AI is stepping into every field of life such as advanced engineering, astronomy, medical science and many more. The perception of attractiveness is different for different individuals. A layperson evaluates attractiveness differently than a dentist and overall, both their perception is different from an orthodontist's point of view. In orthodontics, it is very important to understand the regions which affect the overall facial attractiveness to deliver the best treatment plan to the patient. The aim of this study was to find the difference in perception of attractiveness of the face and smile among the various professional fields. Eye tracking enables us to gather suitable data regarding the visual exploration of the professionals on the respective images shown to them. Since 2010 till date there have been many studies that are focused on evaluating the perception of the beauty of an individual. Various researchers have found that facial attractiveness is the dominating factor above dental attractiveness and others found it to be the opposite. Till today there are not enough evidence on how the perception of beauty is affected by each class of viewers. Richards M. R. et al. mentioned that in a person with malocclusion dental attractiveness captures more attention over background facial attractiveness, whereas in a person with mild malocclusion the visual attention shifts more towards the eye. Richards M. R. et al. and Wang X. et al. also mentioned that the presence of malocclusion highly affected a person's social life by reducing their confidence levels. In our study we have included only females from the north Indian population because firstly, females are more concerned about their aesthetics and are judged more than males. Secondly, the north Indian population was chosen to keep statistical data simple and for ease of data collection. Initially we tried to build our own software for eye tracking by using Open CV but there was a problem during input of multiple images and questionnaire along with images. Further training and time were required to explore working on Open CV. Then it was decided to continue our study on a predesigned online-based eye-tracking tool and Python. We have also found that among all the classes of observers from their respective fields they carry a "Natural Bias" in their way of viewing, e.g., orthodontist on the teeth, endodontist on teeth, periodontist on lips, gums and teeth, for a HR this cannot be defined exactly. For a beautician it is the eyes, for a photographer it is overall face. It was expected at the beginning of the study that these areas are going to receive more amount of fixation duration.

   The questionnaire which was used along with the images helped to collect the subjective data that was quite natural and it matched with the natural bias that we had already mentioned. After analysing all the data from the eye tracking it was found that for a less-attractive image of a full face for all classes of viewers, the attention was towards the eyes, then the mouth and then the nose and hair in a similar level. This value was different for orthodontists and endodontists. For an orthodontist the attention was more on the mouth than the eye, then on the nose. For endodontists, the eye and mouth had almost similar values and then finally their attention fixed on the nose. For the average attractive image all the class of

TABLE 9.2
Overall Fixation Duration of Less Attractive Images

	Overall Fixation Duration of Less Attractive Images				
	Eye	Mouth	Nose	Hair	Chin
Beauticians	1988.4	1166	193.8	578.8	96.7
HR	2351.7	1516.3	123.2	11.4	95.1
Photographers	2841.3	968.7	111	24.5	0
Orthodontist	1829.1	2020.7	226.6	31	172.4
Endodontist	1877.6	1806.1	232.7	33.6	37.2
Periodontist	2076.5	1324.4	165.5	33.5	191.7

viewers' attention was more towards the eyes. Similar values were found for a highly attractive group of viewers.

In the smile image all the class of viewers' attention was towards the teeth irrespective of the level of attractiveness. A trend in the viewing pattern was also detected, i.e., teeth to upper lip and then lower lip.

Intra-region and inter-region reliability was positive and significant in our study as the tracking session was only initiated after a specific value of calibration was achieved, whereas the intra-rater correlation might not match and there is a negative value because the perception among each class is also different depending upon a person's overall psychological views (Tables 9.2–9.14, Figures 9.3–9.14)

The results of our study coincide with the values and the result concluded by Richards M. R. et al. that eyes are the window of attention in each and every class of individual. As mentioned by Johnson et al. the increase in treatment needs from attractive to unattractive the perception shifts from eye towards mouth. It is really necessary to understand the perception of the patient why they require treatment. Eye tracking may be used as an important tool for the evaluation of the need for

TABLE 9.3
Overall Fixation Duration of Average Attractive Images

	Overall Fixation Duration of Average Attractive Images				
	Eye	Mouth	Nose	Hair	Chin
Beauticians	2173.5	678.4	169.1	320.7	27.6
HR	2566.8	777.3	100.4	18.45	0.00
Photographers	2749.8	615.3	123.6	0.00	0.00
Orthodontist	2265.7	1520.1	179.4	0.00	91.8
Endodontist	2163.5	726.6	124.5	12.3	21.5
Periodontist	2379.3	422.1	185.1	0.00	0.00

TABLE 9.4

Overall Fixation Duration of Highly Attractive Images

	Overall Fixation Duration of Highly Attractive Images				
	Eye	Mouth	Nose	Hair	Chin
Beauticians	2756.6	705.9	249	64.1	0.00
HR	3024.7	458.4	150	0.00	0.00
Photographers	2879.8	724.7	72.1	0.00	0.00
Orthodontist	2698.5	263.5	105.6	0.00	72.5
Endodontist	2526	564.4	167.2	0.00	0.00
Periodontist	2606.8	627.2	160.1	0.00	0.00

TABLE 9.5

No. of Fixation Duration of Less Attractive Images

	No. of Fixation Duration of Less Attractive Images				
	Eye	Mouth	Nose	Hair	Chin
Beauticians	4.9	3.6	0.8	1.7	0.2
HR	3.8	3.9	0.3	0.2	0
Photographers	6.7	3.3	1	1	0
Orthodontist	3.2	3.8	1	0.2	1
Endodontist	4.1	5	1	0.2	0
Periodontist	4.6	3.9	1	0.2	1

TABLE 9.6

No. of Duration of Average Attractive Images

	No. of Duration of Average Attractive Images				
	Eye	Mouth	Nose	Hair	Chin
Beauticians	5.7	1.8	0.7	1.1	0.1
HR	6.1	2.5	0.4	0	0
Photographers	5.7	2.6	0.9	0	0
Orthodontist	5.2	3	1	0	0.9
Endodontist	4.9	2.4	1.1	0	0.1
Periodontist	5.2	2.4	1.1	0	0

TABLE 9.7
No. of Fixation Duration of Highly Attractive Images

	No. of Fixation Duration of Highly Attractive Images				
	Eye	Mouth	Nose	Hair	Chin
Beauticians	7.2	1.7	0.9	0.5	0
HR	7.6	2.1	1	0	0
Photographers	6.6	2.5	1	0	0
Orthodontist	7.1	2.3	1	0	1
Endodontist	6.4	2.4	1	0	0
Periodontist	6.7	3.1	1.3	0	0

TABLE 9.8
Overall Fixation Duration of Less Attractive Images (Smile)

	Overall Fixation Duration of Less Attractive Images (Smile)		
	Upperlip	Teeth	Lowerlip
Beauticians	727.4	2913.4	676.8
HR	366.4	3016.6	322
Photographers	409.2	3201	379.3
Orthodontist	758.4	3224.5	367.1
Endodontist	254.4	3372.1	252.1
Periodontist	325.2	3241.5	331.8

TABLE 9.9
Overall Fixation Duration of Average Attractive Images (Smile)

	Overall Fixation Duration of Average Attractive Images (Smile)		
	Upperlip	Teeth	Lowerlip
Beauticians	605.7	3185.2	355.1
HR	405.3	3181	333.9
Photographers	359.6	3001.5	287.9
Orthodontist	482.9	3277.8	520.1
Endodontist	331.2	3180.3	355.6
Periodontist	470.4	3117.4	299.7

TABLE 9.10

Overall Fixation Duration of Highly Attractive Images (Smile)

	Overall Fixation Duration of Highly Attractive Images (Smile)		
	Upperlip	Teeth	Lowerlip
Beauticians	596.8	3166	421.9
HR	282	3083.6	267.4
Photographers	447.8	3106.6	411.6
Orthodontist	794.2	3101.4	271.2
Endodontist	278.1	3236.9	291.1
Periodontist	324.4	3057.9	334.4

TABLE 9.11

No. of Fixation Duration of Less Attractive Images (Smile)

	No. of Fixation Duration of Less Attractive Images (Smile)		
	Upperlip	Teeth	Lowerlip
Beauticians	3.1	7.4	3.2
HR	2.3	6.9	2.1
Photographers	2.3	7.4	3.7
Orthodontist	3.2	7	2.3
Endodontist	2	7.3	2
Periodontist	2.1	7.4	1.9

TABLE 9.12

No. of Fixation Duration of Average Attractive Images

	No. of Fixation Duration of Average Attractive Images		
	Upperlip	Teeth	Lowerlip
Beauticians	2.9	8.1	1.8
HR	2.8	7.4	5.9
Photographers	2.5	7.7	1.6
Orthodontist	2.7	7.5	2.7
Endodontist	2.2	7.9	2.1
Periodontist	2.7	7.5	1.8

TABLE 9.13
No. of Fixation Duration of Highly Attractive Images

	No. of Fixation Duration of Highly Attractive Images		
	Upperlip	Teeth	Lowerlip
Beauticians	3	7.7	2.2
HR	2.8	7.5	2.3
Photographers	2.4	7.2	2.5
Orthodontist	3.2	7.2	2.1
Endodontist	1.9	7.3	4
Periodontist	2.1	6.8	2.1

TABLE 9.14
Intra-region and Inter-Region Reliability

	Intra-region Reliability			Inter-region Reliability		
	ICC	LSB	USB	ICC	LSB	USB
Fixation Duration						
Eye	0.93	0.92	0.94	0.87	0.86	0.88
Mouth	0.89	0.88	0.90	0.84	0.83	0.85
Nose	0.85	0.84	0.85	0.78	0.77	0.78
Hair	0.72	0.71	0.73	0.45	0.43	0.49
Chin	0.77	0.76	0.78	0.55	0.52	0.58

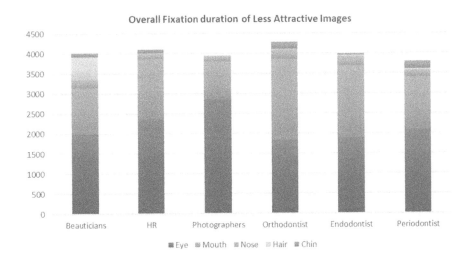

FIGURE 9.3 Overall fixation duration of less-attractive images

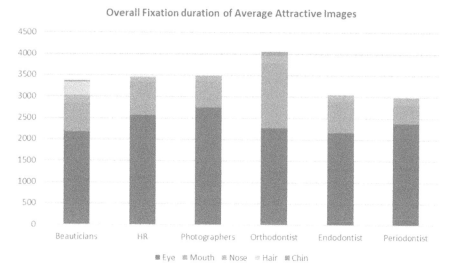

FIGURE 9.4 Overall fixation duration of average attractive images

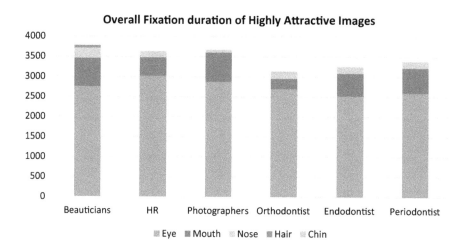

FIGURE 9.5 Overall fixation duration of highly attractive images

treatment in orthodontics. In the future this will help in achieving patients' as well as orthodontists' satisfaction.

CONCLUSION

From this study it was concluded that:

1. Eye tracking is a reliable tool for evaluating facial attractiveness.

No. of fixation (Less Attractive)

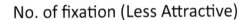
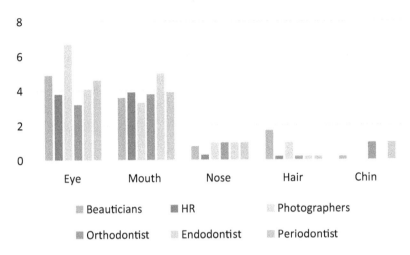

FIGURE 9.6   No. of fixation duration of less attractive images

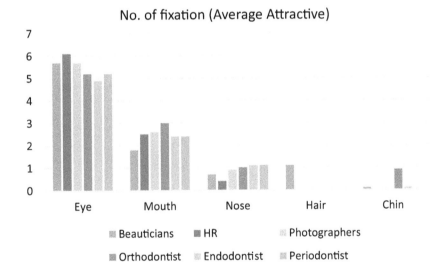

FIGURE 9.7   No. of duration of average attractive images

2. Different classes of viewers view faces differently.
3. Dental professionals are more concerned with the mouth while non-dental professionals are more concerned with the eyes for a less attractive image.
4. Dental attractiveness greatly influences the overall attractiveness of a face in average and less attractive images.
5. It is necessary for a person with a less attractive smile with malocclusion to get an orthodontic treatment to increase their overall attractiveness.

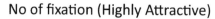
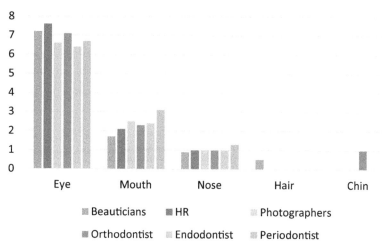

FIGURE 9.8 No. of fixation duration of highly attractive images

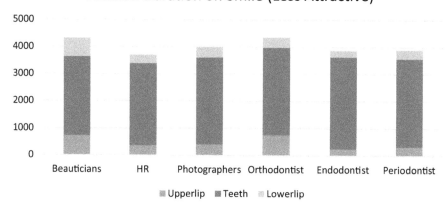

FIGURE 9.9 Overall fixation duration of less attractive images (smile)

Fixation Duration on Smile (Average Attractive)

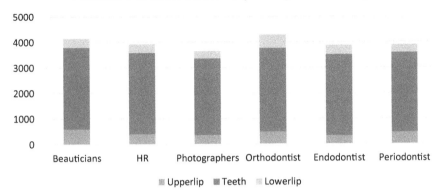

FIGURE 9.10   Overall fixation duration of average attractive images (smile)

Fixation Duration on Smile (Highly Attractive)

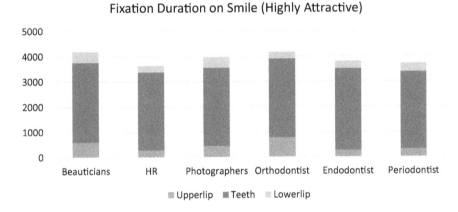

FIGURE 9.11   Overall fixation duration of highly attractive images (smile)

No. of fixation in Smile (Less Attractive)

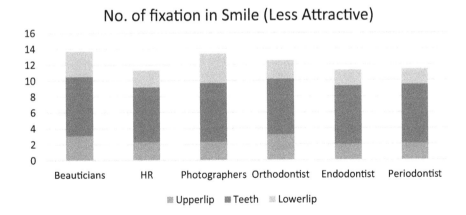

FIGURE 9.12   No. of fixation duration of less attractive images (smile)

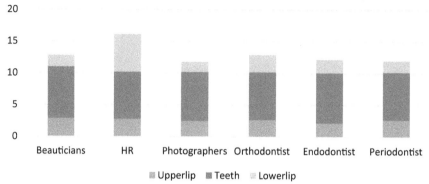

FIGURE 9.13 No. of fixation duration of average attractive images

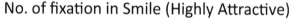

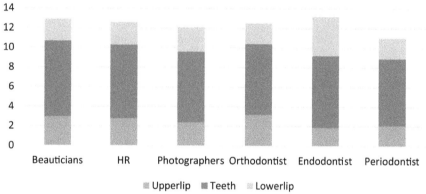

FIGURE 9.14 No. of fixation duration of highly attractive images

REFERENCES

1. Shaw WC, Rees G, Dawe M, Charles CR. The influence of dentofacial appearance on the social attractiveness of young adults. *American Journal of Orthodontics.* 1985 Jan 1;87(1):21–6.
2. Bowman S. Facial aesthetics in orthodontics. *Australian Orthodontic Journal.* 2001 Mar;17(1):17.
3. Havens DC, McNamara Jr JA, Sigler LM, Baccetti T. The role of the posed smile in overall facial esthetics. *The Angle Orthodontist.* 2010 Mar;80(2):322–8.
4. Springer NC, Chang C, Fields HW, Beck FM, Firestone AR, Rosenstiel S, Christensen JC. Smile esthetics from the layperson's perspective. *American Journal of Orthodontics and Dentofacial Orthopedics.* 2011 Jan 1;139(1):e91–101.
5. Thomas M, Reddy R, Jayabharath Reddy B. Perception differences of altered dental esthetics by dental professionals and laypersons. *Indian Journal of Dental Research.* 2011 Mar 1;22(2): 242–7.

6. Cotrim ER, Vasconcelos Júnior ÁV, Haddad AC, Reis SA. Perception of adults' smile esthetics among orthodontists, clinicians and laypeople. *Dental Press Journal of Orthodontics.* 2015 Feb;20(1):40–4.

7. Oliveira PL, Motta AF, Guerra CJ, Mucha JN. Comparison of two scales for evaluation of smile and dental attractiveness. *Dental Press Journal of Orthodontics.* 2015 Apr;20(2):42–8.

8. Luria SM, Strauss MS. Comparison of eye movements over faces in photographic positives and negatives. *Perception.* 2013 Nov;42(11):1134–43.

9. Baker RS, Fields Jr HW, Beck FM, Firestone AR, Rosenstiel SF. Objective assessment of the contribution of dental esthetics and facial attractiveness in men via eye tracking. *American Journal of Orthodontics and Dentofacial Orthopedics.* 2018 Apr 1;153(4):523–33.

10. Richards MR, Fields Jr HW, Beck FM, Firestone AR, Walther DB, Rosenstiel S, Sacksteder JM. Contribution of malocclusion and female facial attractiveness to smile esthetics evaluated by eye tracking. *American Journal of Orthodontics and Dentofacial Orthopedics.* 2015 Apr 1;147(4):472–82.

11. Wang X, Cai B, Cao Y, Zhou C, Yang L, Liu R, Long X, Wang W, Gao D, Bao B. Objective method for evaluating orthodontic treatment from the lay perspective: An eye-tracking study. *American Journal of Orthodontics and Dentofacial Orthopedics.* 2016 Oct 1;150(4):601–10.

12. Celikdelen M, Bicakci AA. Factors affecting smile attractiveness: An eye tracking study. *Journal of Research in Medical and Dental Science.* 2020 Sep;8(6):56–70.

13. de Oliveira WL, Saga AY, Ignácio SA, Justino EJ, Tanaka OM. Comparative study between different groups of esthetic component of the index of orthodontic treatment need and eye tracking. *American Journal of Orthodontics and Dentofacial Orthopedics.* 2019 Jul 1;156(1):67–74.

14. Johnson EK, Fields Jr HW, Beck FM, Firestone AR, Rosenstiel SF. Role of facial attractiveness in patients with slight-to-borderline treatment need according to the Aesthetic Component of the Index of Orthodontic Treatment Need as judged by eye tracking. *American Journal of Orthodontics and Dentofacial Orthopedics.* 2017 Feb 1;151(2):297–310.

15. Kim SH, Hwang S, Hong YJ, Kim JJ, Kim KH, Chung CJ. Visual attention during the evaluation of facial attractiveness is influenced by facial angles and smile. *The Angle Orthodontist.* 2018 May;88(3):329–37.

16. Huang P, Cai B, Zhou C, Wang W, Wang X, Gao D, Bao B. Contribution of the mandible position to the facial profile perception of a female facial profile: An eye-tracking study. *American Journal of Orthodontics and Dentofacial Orthopedics.* 2019 Nov 1;156(5):641–52.

17. Trevisan C, Pithon MM, Meira TM, Miyoshi CS, Saga AY, Tanaka OM. The visual perception and attractiveness of maxillary central incisor abrasion as evaluated via eye-tracking. *European Journal of General Dentistry.* 2019 Jan;8(1):7–12.

18. Janošević PN, Janošević ML, Perović TM, Stojković BB, Stojanović SM. Assessment of smile esthetics and various types of face profiles. *Acta Stomatologica Naissi.* 2020;36(81):2013–21.

19. Wolanski RB, Gasparello GG, Miyoshi CS, Guimarães LK, Saga AY, Tanaka OM. Evaluation of the perception of smile esthetics, in frontal view, with mandibular laterognatism, through the eye-tracking technique. *Journal of Orthodontic Science.* 2020;9.

10 IoT-Based WBAN
A Survey on Classification and Channel Access Techniques

Sachin Kumar, Shivani Sharma, and Pawan Kumar Verma

10.1 INTRODUCTION

Since Internet of Things (IoT) technologies may be used in a variety of medical fields, including real-time Halt Care Monitoring (HCM), patient health care and information management. They play important roles in the monitoring of patients in the modern healthcare (HC) industry. The use of IoT in HCM [1] systems could provide affordable services and possibly lessen the number of patients who need to be hospitalized. Wireless Body Area Network (WBAN) technology is employed in HCM settings as one of the fundamental IoT advancements to monitor patient health. To increase productivity, an HCM system must include both IoT and WBAN technologies [1]. These systems are based on tiny wireless sensor nodes that have been integrated into the human body. Instead of using traditional monitoring, it practically uses automated devices and sensing technologies to recognize the physical characteristics of the victim. From there, it does much computation, processes that computation and mines the information from that computation. There is a lot of interest in employing new wireless technologies to assist remote patient monitoring in a discreet, cost-effective and dependable way, giving patients individualized sustainable services.

WBAN is one such cutting-edge technology that has the potential to greatly enhance the delivery of HC, diagnostic monitoring, disease tracking and associated medical operations. WBAN is created by a wireless network of sensors that are fixed on or in the body to collect physiological data. It is made up of small-power devices that operate inside or all over the human body to provide various services, including those in the medical field.

WBAN is used in the HC industry to compute either (a) bio-kinetic events like body movements, angular motions and acceleration, etc., and (b) external or internal physiological parameters [2] like blood flow measurement, temperature, blood sugar measurement, heart rate measurement, respiratory rate measurement, pulse

DOI: 10.1201/9781003452645-10

rate measurement, electrical activities, metabolic rates measurement, oxygen saturation measurement, etc. One of the intended uses of WBAN is in HC settings where
the conditions of numerous patients are continually and in real-time monitored. To
install a full WBAN in an HCM system, one of the in-process requirements is the
huge number of patients' physiological signs that are wirelessly monitored.

Designing such an application has several difficulties for the software and hardware. Some of these are low cost, accommodating patient care, high energy efficiency and reliable communication by preventing signal collisions between two
sensors and interference from other external wireless devices. This has generated
various research papers that combine WBAN with IoT technology in the literature.
Even though WBAN systems in HCM are very promising, the ability of biomedical
devices to use the communication link effectively presents a significant difficulty
(concerning collisions) when well-organized [3] channel access procedures are not
taken into account. This could have disastrous effects on how well the WBAN perform in terms of system delay, channel utilization efficiency [3] and WBAN throughput. It is hoped that the development of reliable and effective Medium Access Control
(MAC) protocols will address this issue by coordinating and controlling how the
WBAN sensors access the channel for communication. WBAN typically functions
on a single channel. If a WBAN deployment includes a large number of devices, it
is difficult to transmit the data between devices and the CD may result in collisions
which degrade the performance of the system. According to research on MAC protocols, current efforts are concentrated on creating a protocol that may accomplish
efficient data transmission with low power consumption, no channel collision and
no packet loss. The three types of MAC currently in use are Time Division Multiple
Access (TDMA) [4], Carrier-Sense Multiple Access (CSMA) [3] and Hybrid which
includes the TDMA and CSMA protocol to optimize the network efficiency.

As a result, this chapter presents a study of the current technology used in WBAN.
The objective is to give readers a greater comprehension of the WBAN core concepts
in this developing area. We can sum up our contribution to this work as follows:

- To provide a comprehensive study of WBAN, the system model and some
  potential applications of WBAN.
- To highlight the layers of architecture, characteristics, communication
  technologies and system-level challenges in the field of WBAN.
- This study also provides the WBAN channel access issues, and a survey of
  already proposed MAC protocols to resolve this issue.

10.2 GENERAL SYSTEM MODEL FOR WBAN

The Body Area Network (BAN) [5] is defined by standards such as IEEE 802.15.4
[1] and IEEE 802.15.6 as a logical set made up of nodes (sensors and devices) and a
single CD. It utilizes a star topology-based network with two distinct communication
methods: straightforwardly extended two-hop and one-hop star topologies. A basic
One-Hop Star Topology (OHST)-based network allows for direct frame exchanges
between nodes and the BAN hub but extended two-hop topologies incorporate a

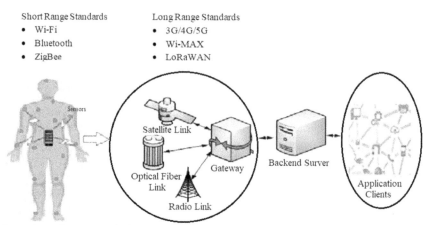

Short Range Standards
- Wi-Fi
- Bluetooth
- ZigBee

Long Range Standards
- 3G/4G/5G
- Wi-MAX
- LoRaWAN

Intra-WBAN Tier-1 Communication Inter-WBAN Tier-2 Communication Beyond WBAN Tier-3 Communication

FIGURE 10.1 General system model for WBAN

relay device and allow for direct or indirect transmission between devices and the hub, as depicted in Figure 10.1.

The WBAN communications design typically consists of three parts: intra-WBAN Tier-1 communications, inter-WBAN Tier-2 communications and beyond-WBAN Tier-3 communications.

10.2.1 Intra-WBAN Tier-1 Communication

The term *"intra-WBAN communications"* refer to wireless communications occurring within 2 meters of a person's body, and it is further divided into connections between body sensors and portable devices. The direct connection between the WBAN and the body area sensors makes the intra-WBAN communications' architecture critical. The initial tier of data communication covers between intra-sensor nodes with the CD. In this level of communication, the sender and the receiver are both inside the body range. Off-body, on-body and in-body sensors (nodes) are included in this. In this communication, the properties of the physical layer, sensor nodes, MAC layer [6] and the frequency used all affect the data rate.

10.2.2 Inter-WBAN Tier-2 Communication

The data transmission between two separate WBANs and the data transmission between the hubs and the CD such as Access Points (APs) or coordinators are both considered to be Tier-2 communications. In contrast to Wireless Sensor Networks (WSN), which frequently function as independent systems, the WBAN is rarely used alone. When handling WBAN emergencies, the CD can be regarded as one of the key components of the infrastructure of the dynamic Tier-2 environment. The Tier-2 WABN network feature is accordingly used to connect the BANs to different (easily available) networks, including the cellular and Internet networks. The following categories describe the inter-WBAN communication paradigms: ad hoc-based building, which enables quick sharing in dynamic circumstances such as medical emergency response scenarios and WBAN infrastructure-based development, which provides huge bandwidth with centralized control and flexibility. Architecture based on WBAN limited space applications, including those used in workplace, home and hospital settings heavily rely on infrastructure-based and inter-WBAN communications. The infrastructure-based networks enable centralized administration and security supervision. Additionally, the CD can serve as a database server for certain applications like SMART and CareNet [6].

10.2.3 Beyond WBAN Tier-3 Communication

All communications that occur outside of the WBAN fall under the category of Tier-3 communications. Internet-based communication between medical backend servers and CD is part of Tier-3. The TCP/IP stack has comprehensive definitions for each protocol used at this level of communication [6].

10.3 WBAN LAYERS' COMMUNICATION STRUCTURE

Generally, all 802.15.x permitted standards propose both the MAC layers and PHYS (Physical Layer). The IEEE 802.15.6-based WBAN vigorous collection offers extremely low power consumption, great reliability, low cost and minimal complexity. The network management data are typically connected to the PHYS by a logical node management entity and CD management entity [7].

10.3.1 Physical Layer (PHYS)

The deactivation and activation of the WBAN transceiver, the transmission and receipt of data and Clear Channel Assessment (CCA) [7] in the existing channel are all functions that fall under the purview of the PHYS layer. The choice of PHYS is depending on the context at hand, including non-medical and medical as well as over-body, outside-body, and in-body devices communication. Radio transceivers (WBAN) are used to start and stop data transmission and reception. The CCA in the current channel is all under the control of the PHYS. The PHYS Service Data Unit (PSDU) can be converted into a PHYS Data Unit (PPDU) using a method provided by the PHYS layer. According to the NB standards, the PHYS Preamble (PLCP) and PHYS Header (PSDU) should be pre-attached to the PSDU to create PPDU. The PCLP header is then sent through the data communication rates indicated in its effective bandwidth after the PCLP Preamble. The PSDU is the final PPDU module and includes a Frame Check Sequence (FCS) and MAC-header/frame body. The Start Frame Delimiter (SFD)/including preamble, packet format and modulation, are provided by the PHYS. Only time and the preamble series are sent four times throughout the SFD series to guarantee packet synchronisation [8]. Payload length, pilot data, synchronization, data rate and WBAN ID constructed over the PHYS Header are all included in the PHYS Header [8].

10.3.2 MAC LAYER

The MAC layer is defined based on the IEEE 802.15.4 [1] or IEEE 802.15.6 working assembly on the upper portion of the PHYS to govern the channel access. For the purpose of allocating time reference resources, the CD divides the time axis [9] or the complete channel into the chain of superframes (SPs). To bind the SPs, it chooses beacon periods of identical length. One of the following channel access modalities is used by the CD for channel access coordination:

(1) Beacon period and beacon mode SF boundaries [9]: The beacon is used to synchronize the nodes in the network. The CD directs beacons during each beacon period unless SPs are dormant. The CDs use beacon frames or timed frames to control the SP structure communication.
(2) Mode with SP borders and non-beacon enable: It is unable to broadcast beacons. It is compelled to use the SP structure's time frames.
(3) Mode without SP borders and non-beacon enable: The CD only provides unplanned Type II polled allocation in this mode. Each node must therefore separately decide on its timetable [9].

The following access methods are available throughout each superframe period:

(a) Contention-free access that is connection-oriented: It plans the slot distribution across one or more upcoming superframes.
(b) Access without a connection or interruption: For resource distribution, it makes use of posting and polling.
(c) Random access algorithms: For resource allocation, it either employs the Carrier Sense Multiple Access with Collision Avoidance (CSMA/CA) [1] or the slotted ALOHA technique.

10.4 APPLICATIONS OF WBAN

Non-medical and medical WBAN applications are separated by IEEE 802.15.6 or IEEE 802.15.4. The primary aim of all WBAN applications is to improve the quality of life for the user. WBAN's technical needs, however, depend on the application. Figure 10.2 displays a few on-body and in-body applications [10].

10.4.1 HEALTH CARE (HC) APPLICATIONS

By identifying numerous life-threatening disorders and offering immediate patient treatment and monitoring, WBAN has tremendous potential to alter the future of HC treatment and monitoring. In 2025, the global population is expected to have doubled from its 1990 level of 357 million, according to demographers. This suggests that by

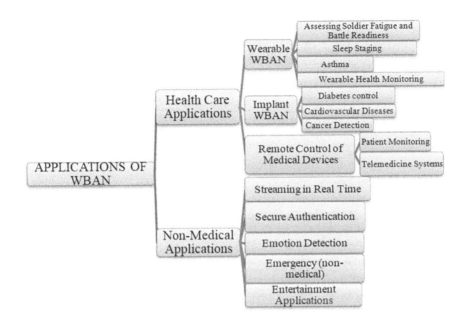

FIGURE 10.2 On-body and in-body applications of WBAN

the year 2050, medical elderly care will pose a significant problem. The cost of HC in the United States (US) [10] was over 2.91 trillion in 2009, and it is predicted to rise to 5.1 trillion by 2025, or nearly 20% of GDP. Additionally, cardiovascular disease, which accounts for up to 30% of fatalities globally, is one of the main causes of death. With the technological advancements in sensor integration and miniaturization, the Internet [10] and wireless communication, the distribution and service of HC will be radically reorganized. The adoption of WBAN is anticipated to improve HC systems by enabling better disease management, illness diagnosis and crisis response rather than only well-being. Ongoing observation of one's physiological traits such as heart rate, blood pressure [10] and body temperature, is possible because of the use of WBAN in medical applications. Data acquired by the sensors can be communicated to a CD, such as a mobile phone when abnormal conditions are identified. The gateway then transmits its data to distant places, such as a critical care department or a doctor's office, through a cellular or PAN or the Internet so that a decision can be made. Additionally, monitoring, early detection and treatment of individuals with potentially serious diseases of numerous types, such as hypertension, diabetes and cardiovascular-associated ailments, will be made possible by WBAN. WBAN can be used for three different types of medical purposes, which are as follows.

10.4.1.1 Wearable WBAN

The following two subcategories of wearable HC applications of WBAN are further categorized as follows: (a) disability help and (b) management of person performance. Several of these applications are listed below [10].

10.4.1.1.1 Evaluating the Combat Readiness and Fatigue of Soldiers

WBAN can keep a closer eye on how soldiers are behaving on the battlefield. A WBAN with cameras, wireless networking, biometric sensors and an aggregation gadget for sharing information with centralized monitoring and other soldiers can do this. A secure data transmission channel between the soldiers is necessary to prevent ambushes. While offering better monitoring and taking precautions in case of injury, the usage of WBAN in hostile environments can be crucial in lowering the risk of damage [10].

10.4.1.1.2 Supporting Professional and Recreational Sport Training

WBAN makes it simple to adjust an athlete's training regimen because they offer motion capture, monitoring parameters and recuperation. Additionally, the user-provided real-time feedback in these networks enables performance enhancement and minimizes accidents brought on by improper training [10].

10.4.1.1.3 Insomnia Staging

Sleep is a crucial behavioural and regular physiological process that takes up one-third of our waking hours. Approximately 27% of people worldwide (on average) experience sleeps disorders. Such problems can have very serious side effects, including heart disease, workplace sleepiness and drowsy driving. The cost of lost production due to sleep disorders is estimated at US$18 billion. As a result, recently, there has been a lot of interest in sleep monitoring. A polysomnography examination, which includes an examination of various biopotentials recorded over the course of a night in a sleep laboratory, can identify sleep disorders. The patient is prevented from falling asleep by the several cables that must be run from the head to a device fastened to the patient's belt for these measurements. Additionally, it impedes the patient's movements and starts noise and artifacts that degrade the signal superiority. The cleverness and tools in WBANs' sensor nodes can be delocalized, and all wires can be cut [10].

10.4.1.1.4 Asthma

The associated sensors in WBAN can monitor allergic chemicals in the atmosphere and give a doctor instantaneous feedback, which could advantage millions of patients [10].

10.4.1.1.5 Health Monitoring through Wearable Devices

WBAN can provide immediate health monitoring when used with health devices and sensors on the person's body. For instance, patients with diabetes can use a glycocoll phone with a glucose module. The glucose module sends glucose diagnoses to the cell phone, which can subsequently store or send the diagnosis to a physician for analysis [10].

10.4.1.2 Implant WBAN

This category of WNAN applications relates to sensors inserted into the body of a person, either subcutaneously or intravenously. The following are some of these applications.

10.4.1.2.1   Diabetes Control

In 2010, 286 million adults worldwide, or 6.41% of the grown population, had diabetes. By 2031, this figure is projected to increase to 439 million, or 7.81% of the grown population. If diabetes is not thoroughly monitored and treated, research has shown that it can lead to long-term medical problems. Regular monitoring from WBAN can lower the danger of fainting, allow for optimal dose, remove hazards of loss of circulation, prevent blindness later in life and prevent other consequences [10].

10.4.1.2.2   Cardiovascular Diseases

With suitable HC measures, cardiovascular illnesses are recognized to be the leading cause of mortality for 17 million people. By using WBAN technology to track episodic episodes and other aberrant situations, Myocardial Infarction can be significantly decreased.

10.4.1.2.3   Cancer Detection

By 2020, the number of cancer deaths is predicted to rise by 50%, reaching up to 15 million. With the help of WBAN-based sensors that can track cancer cells inside the human body, doctors will be able to continuously identify tumours without a biopsy, allowing for quicker analysis and treatment [10].

10.4.1.3   Medical Devices Remote Control

The *Ambient-Assisted Living (AAL)* network of devices and services is made possible by the pervasive Internet connectivity of WBAN, and each WBAN wirelessly connects to a back-end HC server. AAL strives to prolong the patient's ability to manage their own care while receiving assistance at home, reducing the need for extensive individual care, improving excellence of life and lowering societal expenditures. A new production of IT systems with traits like user-friendliness, context awareness, anticipatory behaviour and adaptability will be encouraged by ambient-assisted living [10].

10.4.1.3.1   Patient Monitoring

Monitoring crucial signals and providing immediate feedback and details on the healing development in HC applications are two important uses of WBAN. They detect and send vital signal measures such as body temperature, heart rate, respiration rate, body implant characteristics, blood pressure and chest noises to be more precise. WBAN can also be used to administer medications in medical facilities, monitor physiological data remotely, support rehabilitation and act as a diagnostic interface. Because WBAN can supply interconnection between a variety of devices on or in the body, such as hearing aids, and other devices, it has potential applications that go beyond patient monitoring, such as trauma care, post-treatment follow-up and remote help in accidents, pharmaceutical research and research into chronic diseases.

10.4.1.3.2   Telemedicine Systems

Existing telemedicine mechanisms either use a less-power-hungry technology like ZigBee, which is susceptible to nosiness from additional devices operating in an analogous frequency, or specialized wireless channels for sending data to distant stations. As a result, they limit extended monitoring. In contrast, incorporating WBAN into a telemedicine system [10] enables continuous, undetectable ambulatory health monitoring.

10.4.2 NON-MEDICAL APPLICATIONS

The following five subcategories further divide non-medical WBAN applications.

10.4.2.1 Streaming in Real Time

This category of applications includes video-streaming features including video clip capture using a mobile phone's camera, trade exhibitions for sporting goods along with the newest trends in 3D and other videos. Additionally, voice communication for headsets enables audio streaming for purposes like listening to art explanations in museums or bus timetable information at bus stops, multicasting for discussion calls and perusing samples of music in music stores. Stream transfer is a subcategory that also includes body information and essential sign-based entertainment services, emotion detection, identification and monitoring of forgotten items by alerting the owner. These applications include entertainment devices' remote control, body motion recognition, identification and stream transfer [10].

10.4.2.2 Entertainment Applications

Social networking and gaming apps are included in this category. Devices that can be utilized as WBAN devices include MP4 players, microphones, digital cameras, cutting-edge computer-based appliances and head-mounted displays. They can be utilized for virtual reality, gaming (including hand gesture controls, movable body motion games and virtual games), personal item monitoring, digital ID card and profile exchange and consumer electronics.

10.4.2.3 Non-medical Emergency

Non-body sensors, such as those used in the home, are capable of spotting non-medical emergencies like fires or toxic/flammable gas leaks, and they must quickly transmit this information to CDs to alert the wearer of the urgent situation [10].

10.4.2.4 Emotion Detection

Recent studies have demonstrated how effectively human emotions may be realized through visual and speech signal data examination. More specially, wearable device sensing technologies have made it possible to detect emotions by inducing physical manifestations all over the body, which produces signals that can be detected by basic biosensors. For instance, panic speeds up the heartbeat and respiration, which causes perspiration on the palms and other symptoms. Therefore, by keeping an eye on physiological signals connected to emotions like the Electrocardiograph (ECG), Electromyograph (EMG), Electroencephalograph (EEG), Electro dermal Activity (EDA), etc., one can monitor their emotional state at any time and anywhere.

10.4.2.5 Secure Authentication

The use of behavioural and physical biometrics, including face patterns, fingerprints and iris recognition is used by secure authentication applications. Due to copy ability and fraud, this is one of the main uses for WBAN, which has inspired the introduction of new behavioural and physical traits of the person's body, primarily gait analysis, electroencephalography and multiple biometrics [10].

10.5 CHARACTERISTICS OF WBAN

Characteristics of WBAN can be categorized as follows.

10.5.1 WBAN NODE TYPES

A WBAN node is a standalone device having communication capabilities. WBAN is body-centric and consists of two types of nodes: CDs mounted on the human body and in-body, on-body, or around-body sensor nodes [11]. Thus, various information about the human body can be gathered and transferred to the CD for assessment and presentation. Data can be forwarded to distant servers by the CD using a wireless and broadband network. As a result, body area data, including vital signs, motions and environmental conditions, can be gathered without interfering with people's daily activities. The many WBAN nodes can be categorized as follows.

10.5.1.1 Sensor

The sensor devices are tiny, have a tiny battery and have little computation, communication and power resources. WBAN uses sensors to measure various body characteristics either externally or internally. These sensors gather data based on physical agitation, react to that data by processing it and transmit the gathered wireless responses to CD [11]. These sensors can be biogenetic, physiological or environmental sensors. Some of the current varieties of these sensors can be included in a person's watch, phone or earbuds, enabling wireless monitoring of that person wherever they are and with whomever. The following are the main elements of the sensor nodes.

10.5.1.1.1 Microcontroller
It executes local data processing, including data compression, and regulates the operation of the other parts.

10.5.1.1.2 Memory
The detected data acquired from the sensor nodes is temporally stored in memory.

10.5.1.1.3 Radio Transceiver
It connects the nodes and enables wireless transmission and reception of physiological data.

10.5.1.1.4   Power supply
It is utilized to provide the sensor nodes with the necessary battery-powered power.

10.5.1.1.5   Signal Conditioning
The physiological detected data are filtered and amplified to levels appropriate for digitalization.

10.5.1.1.6   Analogue to Digital Converter
To enable additional essential procedures, it converts analogue signals to digital ones.

10.5.1.2   Actuator
After receiving data from the sensors, the actuator engages with the user. By responding to sensor data, it serves as a network feedback mechanism; for instance, by injecting the proper dosage of medication into the person's body in ubiquitous HC applications. According to how they are implemented within the body [1], IEEE 802.15.4 has suggested another categorization for nodes [11] in a WBAN, which is given as follows.

10.5.1.2.1   Implant Node
Either directly beneath the epidermis or within biological tissue, this kind of node is implanted within the human body.

10.5.1.2.2   Body Surface Node
In the human body, this kind of node is implanted either directly beneath the skin or into bodily tissue.

10.5.1.2.3   External Node
This kind of sensor node is placed from a few millimetres to a few meters outside the person's body, not in contact with it.

10.5.2   A WBAN's Nodes Number

A WBAN's nodes number is reported to vary from a few sensors, smart meters or actuators communicating [11] with a moveable handset up to one to hundreds of actuators, smart meters or sensors connecting with an Internet gateway, which are draughts of IEEE standards associated to the technical necessities of WBAN. According to reports, a typical WBAN-based medical network has six nodes and can scale to handle up to 256 nodes. Due to restrictions on the transmission technique, only one CD is permitted to communicate in a WBAN, with the IEEE 802.15.6 specification having nodes ranging from 1 to 64.

10.5.3   Topology Used in WBAN

According to the IEEE 802.15.6 operational group, either a one-hop or two-hop [11] star topology can be used by WBAN, with the nodes in the star's core situated somewhere like the waist. The OHST allows for the transfer of two types of data [11]:

from the sensors to the CD and from the CD to the sensors. In the star topology [11], communication channels are accessible in both non-beacon and beacon modes. The device at the centre of the network is known as the CD, manages communication in the beacon mode. To support network organization control and sensor synchronization, it periodically emits beacons that specify the beginning [2] and finish of a SP. The user can specify the duty cycle of the communication network, which is the duration of the beacon interval, and it depends on the WBAN standard. When operating in non-beacon mode, devices in the WBAN network can send information to the CD by using CSMA/CA. To receive data, the nodes must be powered on and polled by the CD. However, the CD is unable to always connect with the nodes because they have to wait to be requested to join a communication. Given that WBAN can use both one-hop and two-hop [11] star topologies, significant considerations regarding power are necessary. Relays are used to spread the transmission heat and provide a comfortable level of heat for the sensor's immediate surroundings [11].

10.6   COMMUNICATION TECHNOLOGY

As shown in Figure 10.1, a typical WBAN must execute two different wireless communications tasks for the incorporation of prospective IoT-based applications, data transmission from a wearable to a CD and data transmission from a CD to the Internet cloud [12]. Table 10.1 shows various communication protocols used at different communication layers in WBAN scenario.

To communicate the data to a CD in the first scenario, wearable sensor nodes often create a wireless network with a simple topological configuration (e.g., a star topology). For the latter, data are typically transmitted between a CD and the Internet-based cloud via long-range wireless communication technologies. Although there are many wireless communication technologies that are currently accessible, including WLAN, Bluetooth, 4G, LTE, UMTS, Wi-Fi, ZigBee and WiMAX. The examined WBAN primarily employed Wi-Fi to accomplish it. Short Range Communication (SRC) technology is used for short-range Inter-WBAN Tier-1 communication, and Long-Range Communication (LRC) technology is used for long-range Inter-WBAN Tier-2 communication. Both types of technology can be employed in WABN [12].

TABLE 10.1
WBAN Layers Communication Structure [1], [4]

(a) Communication Layers	(b) Protocols Used
Application Layer	CoAP, AMQP, DDS, MQTT-SN, HTTP, XMPP, REST
Network Layer	IEEE 802.11, IPv4/IPv6, IEEE802.15.4, 6LoWPAN
Data Link/MAC Layer	PCA, CSMA, CSMA/CD, CSMA/CA, ALOHA, SLOTTED ALOHA,
Physical Layer	LTE-A, IEEE802.15.4, BLE, Z-Wave

10.6.1 SRC TECHNOLOGIES

The SRC technology is used for a short distance, typically for Inter-BAN Tier-1 communication. Typical examples of SRC technology are Bluetooth, Wi-Fi, ZigBee, etc.

10.6.1.1 Bluetooth

The most popular wireless communication standard utilized for information transmission from sensors to a CD is IEEE 802.15.1 (Bluetooth) [12]. Bluetooth uses a 2.4 GHz license-free spectrum for data transmission. It is a cost-effective and low-power specification used for SRC, particularly in moveable devices. Typically, it has a range of about 10 meters and provides data rates of 721.21 kbps for Bluetooth model [12] and 3 MB/s for Enhanced Data Rate. The majority of WBAN in this technology utilized Bluetooth Low Energy (BLE), a very low-power variant of the Bluetooth protocol. It incorporates the Wi-Fi PHYS/MAC layers for faster throughput and supports a throughput of 1 Mb/s with an utmost communication range of larger than 66 m.

10.6.1.2 Wi-Fi

Wi-Fi is the most frequently used communication technology in the studied literature. Wi-Fi is a term used in the industry to refer to a particular type of WLAN protocol that depends upon the network standard of IEEE 802.11x. Depending on the kind, it operates on the 5 GHz or 2.4 GHz radio band and may transmit data at rates between 11 Mb/s and 450 Mb/s. Its communication range typically ranges from 30 m to 250 m [4], based on the type and requirement conditions.

10.6.1.3 ZigBee

Many WBANs also employ ZigBee communication technology. A Wireless Personal Area Network (WPAN) [13] standard identified as ZigBee defines both the application, network and MAC layers. It was first included in the official standard of IEEE 802.15.4. With a longer battery life requirement, this technology aims for short data rate, short distance, low complexity and low-cost solutions. It may operate in 1, 10 or 16 channels with data rates (maximum) of 20 kbps, 40 kbps or 250 kbps accordingly using radio airwaves at 868/915 MHz or 2.4 GHz. With the highest communication range of around 75 meters, ZigBee devices can communicate using star, tree or mesh topologies [13].

10.6.2 LRC TECHNOLOGIES

The LRC technology is used for long-range communication typically for Inter-BAN Tier-2 communication. Typical examples of LRC technology are *WiMAX*, *LoRaWAN*, SIGFOX, *NB-IoT*, etc.

10.6.2.1 WiMAX

The WiMAX (worldwide interoperability for microwave access) [13] technology described by the IEEE 802.16 standard enables point-to-multi-point wireless broadband access. The gaps between WAN and LAN technologies have been filled by

WiMAX technology. WiMAX technology offers high bandwidth connectivity at a low cost, with broad coverage and lots of users. In the MAN network, fixed broadband access is provided by the IEEE 802.16d standard, and mobile broadband access is provided by IEEE 802.16e technology. High data rates, good spectrum efficiency and a strong communication range of up to 50 km are all provided by WiMAX technology [2].

10.6.2.2 LoRaWAN

An energy-efficient long-range wireless communication technology called LoRaWAN is utilized to link up WBAN devices that run on batteries [12]. Semtech has trademarked the spread spectrum modulation method known as LoRa. It can run on 868/915 MHz [13] with a maximum data rate of 50 kb/s. Applications that require data transfer over large distances at small data rates and transmission frequencies can use this technology.

10.6.2.3 SIGFOX

SIGFOX offers low-power communication technology for WBAN on the 433 MHz, 868 MHz and 915 MHz frequencies in Asia, Europe and North America, respectively. Up to a communication range of 10 km and 40 km, SIGFOX offers data rates of 100 bps [13]. To increase receiver sensitivity and reduce energy usage, SIGFOX employs Ultra Narrowband (UNB) modulation techniques. Devices using SIGFOX technology are capable of sending up to 140 messages per day [13]. Utilizing a 200 kHz macro channel in the ISM band, the SIGFOX MAC layer transmits three messages [13].

10.6.2.4 NB-IoT

Narrowband IoT is a low-power WAN cellular technology created by 3GPP (NB-IoT). For IoT networks, NB-IoT employs LTE standards and offers low cost, low power and a sizable area [14]. NB-IoT technology [14] uses single-carrier frequency division multiple access techniques to provide 250 kbps uplink throughput and 230 kbps downlink speed while connecting a sizable number of devices (50,000). Up to 35 kilometres can be covered using NB-IoT. Twelve subcarriers spread across 180 kHz of bandwidth are used by NB-IoT technology (15 kHz each). Single-tone and multi-tone transmission are the two forms of transmission offered by NB-IoT [14].

10.7 WBAN'S CHALLENGES AND OPEN ISSUES

To implement the WBAN, the following are the primary obstacles and open problems:

10.7.1 ENVIRONMENTAL CHALLENGES

The substantial path loss caused by body absorption in the WBAN is a problem. Multi-hop links and heterogeneous networks with various sensors placed in various locations should be used to reduce this. The complexity of the channel models increases as a result of multi-path and mobility. Additionally, due to various WBAN limits regarding the antenna material, shape [15], size and different RF requirements, antenna design presents more difficult problems. In actuality, the antenna's placement is determined by the implant [15].

10.7.2 PHYSICAL LAYER CHALLENGES

To reduce power consumption without sacrificing dependability, PHYS layer protocols must be established. These techniques need to work well in areas that are susceptible to interference. Advances in low-power RF technology can reduce maximum power consumption, resulting in disposable patches with low production volumes. The peak power consumption of the WBAN should be 0.001–0.1 mW [16] in reserve mode and up to 30 mW [16] in fully active mode, and they should be scalable. One of the major shortcomings of WBAN is interference, according to research. When people wearing WBAN devices enter each other's range, it happens. With greater WBAN density, the coexistence problems are more obvious. Additionally, erratic postural movements of the body may help the networks move into and out of each other's field of view.

10.7.3 MAC LAYER CHALLENGES

The MAC protocol as a whole is not constructed by the IEEE 802.15.6 procedures. Only the fundamental conditions are needed to guarantee device interoperability for IEEE 802.15.6 MAC. The MAC protocols should also enable the extension of the lifetime of the sensor, WBAN applications' need for energy efficiency, the reduction of energy consumption [17] and flexible duty cycling. In general, the MAC protocols for WBAN do not provide adequate network throughput. Such protocols cause performance delays in fluctuating traffic conditions and difficult power needs. Additionally, the WBAN have specific Quality of Service (QoS) requirements that should be met by the MAC.

10.7.4 TRANSPORT CHALLENGES (QoS)

Applications for WBAN should be able to satisfy QoS criteria without experiencing performance loss or an increase in complexity. The limited memory of WBAN necessitates effective retransmission, error detection techniques and secure rectification. Smartphones are now more widely accepted, ubiquitous and powerful. As a result, carrying a current wireless device with several functions has become more acceptable, thanks to mobile health (mHealth) technologies. Smartphones can also be utilized in unusual ways to track sleep. The job of WBAN and huge data are made clear by the conversation that came before it. Both methods play a part in various applications, particularly in the field of medicine [3].

10.8 PERFORMANCE METRICS FOR IOT-BASED WBAN

A network's performance is based on how the user perceives the quality of its services. There are numerous methods to assess the performance of the network depending on its design and kind. Indicators of network performance include jitter, latency (delay), bandwidth and throughput.

10.8.1 BANDWIDTH

The network's bandwidth allocation is one of the significant factors affecting how well a network performs. The amount of information or data that may be transmitted in a particular period is known as bandwidth [18]. Bandwidth has two separate estimation values and can be applied in two different circumstances. Bits per second (bps) [19] is used to assess bandwidth for digital devices. The bandwidth of analogue devices is expressed in cycles per second, or Hz (Hz). One element in deciding what a user perceives as a network's speed is bandwidth.

10.8.2 THROUGHPUT

The throughput is the amount of effectively transferred data (packets) per unit of time. Hardware restrictions, accessible bandwidth and signal-to-noise ratio all have an impact on throughput. Consequently, a network's top throughput could be better than the experienced throughput during normal usage. Although they are not the same, throughput and bandwidth are frequently confused. As opposed to bandwidth, which represents the maximum capacity (potential) of a network, throughput is the real capacity of how rapidly we can transmit data [18]. Throughput is influenced by various factors, together with the resistance of the underlying analogue physical media, the processing power provided by network components [19], and end-user behaviour. When various protocol operating expenses are taken into consideration, the utilization rate of the supplied data may be significantly lesser than the utmost throughput feasible.

10.8.3 LATENCY

The time taken by the network to deliver data completely from the sender (source) to the receiver (destination) through a network, starting at the moment the initial bit of the data are sent out and ending at the moment the final bit of the data are received at the receiver is known as latency (also known as delay). When a network connection is in a low-latency network, there are only minor delays, but when a network connection is in a high-latency network, there are large delays. Any communication over a network that has a lot of latency experiences bottlenecks; by restricting data from utilizing the WBAN channel to its full potential, significantly reduced the bandwidth of the communication WBAN. The consequence of latency on a WABN's capacity might be either temporary or permanent depending on what caused the delays [20]. Latency is also recognized as ping rate [20] a millisecond-based unit of measurement (ms).

10.8.4 JITTER

Another performance problem associated with delay is jitter. Jitter is a "packet delay variance" in computer communication networking. Jitter is merely referred to as a difficulty when several data packets in a network have dissimilar delays and the packets at the destination is time-critical, such as video data [20]. Jitter is calculated in milliseconds. Jitter is defined as a variation of the regular flow of packet transmission. In other terms, any variation or displacement of the high frequency pulses in a digital stream is known as a jitter. The variance may be associated with the phase timing, signal amplitude and pulse width. Crosstalk and Electromagnetic Interference (EMI) between transmitted signals are the major sources of jitter [20]. Jitter can affect a monitor display to flicker, impair the capability of a desktop or CPU to function as intended, set up clicks or other undesirable effects in transmissions and cause the loss of sent data across the network.

10.9 MAC PROTOCOLS AND THEIR REQUIREMENTS FOR WBAN

WBANs are mostly used in HC applications to manage critical care scenarios. In essence, WBAN provides a collection of wireless sensor nodes that exchange medical data from multiple sources to one or more destinations using various physical communication channels. To regulate transmission between the components of a WSN, MAC protocols are created. Since of this, MAC protocol design is crucial because it has a big impact on how effectively WBAN communicates. Therefore, specific design criteria should be met by the MAC protocols created for WBAN to handle requirements including dependability, latency, scalability, energy efficiency and traffic priority [21]. The current MAC protocols deal with some of the above-mentioned features' problems, but they fall short on others. The major MAC protocols of the current approaches are highlighted in the section that follows:

10.9.1 CSMA-BASED PROTOCOLS

This group of MAC protocols uses contention-based methods. No channel has any time slots set aside for it. In this mechanism, prior to data communication, the network nodes watch the channel. The node sends data when it discovers the channel is empty. In general, networks with variable traffic loads are suited for this class of protocols. By evenly allocating time slots for each node, this method can reduce the amount of waste of information in idle time [21].

When data are not given a fixed time frame for delivering requests, contention-based protocols are employed. These protocols are extremely helpful when data transmission is random as compared to periodic data. Contention-based schemes are fundamentally different from scheduled systems since a transmitting user's success is not assured. Due to their simplicity, flexibility and resilience, contention-based protocols like CSMA/CA are frequently employed in wireless sensor networks. To use the channel and convey their data, global topology or synchronization information is not required by the nodes. The node can enter and exit the network without experiencing any serious difficulties. Comparing the contention protocol to scheduling protocols, there are various benefits. Contention protocols can more easily detect changes in traffic load or node density since they allocate resources on demand. Second, as topologies alter, contention protocols may be more adaptable. In this, peer-to-peer data transmission is directly supported, and communication clusters are not required. Finally, unlike TDMA protocols, contention-based protocols do not demand precise time synchronization. A contention protocol's primary drawback is its wasteful use of energy [22].

10.9.2 TDMA-BASED PROTOCOLS

In TDMA protocols, the time at which a node or device can transfer the data is chosen by a scheduling protocol. This allows numerous nodes to communicate concurrently without interfering with each other on the broadcast channel [22]. Typically, the time is split into slots, and those slots are then arranged into data frames. A device is given a minimum of one slot to broadcast throughout each frame. The smallest frame is typically found via a scheduling method to maximize spatial reuse and minimize packet latency. By distributing the data into various slots, TDMA provides multiple devices to share the same frequency channel. It benefits naturally from collision-free media access. A device only has to turn on its communication radio through the slot to transmit or receive because it supports low-duty cycle operation. Node synchronization and topology change adaption are TDMA systems' limitations. As a result, the slot allocations need to take these possibilities into account. However, it is challenging to alter the assigned slot for a typical TDMA in a decentralized system because all devices must be in agreement on the assigned slot. On the foundation of this contention-free channel access, several protocols were suggested. To prevent packet collisions and improve communication reliability, each user is given non-overlapping slots. This prevents data collision in the networks; however, these protocols' performance is limited by issues like scalability and fixed data rates [22].

10.9.3 HYBRID PROTOCOLS

The schedule-based and contention-based MAC protocols are combined to form the hybrid protocol. With the intention of incorporating the advantages of both TDMA and CSMA protocols, various MAC protocols have been developed. The duty cycles (active or sleep) are used by these hybrid procedures, which include segmenting time into data frames, through which a device spends some of its time communicating and sleeps during its downtime to minimize energy loss from idle listening.

Researchers have recently suggested a number of MAC mechanisms for WBAN using various methodologies. Scheduling-based (collision-free), contention-based and Hybrid MAC protocols have all been examined in terms of dependability, latency, quality of service and energy efficiency. Even while certain designs may perform well in terms of energy economy and delay, there are still several obstacles to overcome for real-time service support. The Hybrid MAC protocols demonstrate stronger and more effective capabilities for applications (real-time), but there are still numerous obstacles in sensor networks that need to be overcome before a good solution for real-time energy-efficient communication can be found [22].

10.9.4 WBAN's MAC Protocol Requirements

Energy efficiency, latency, deployment, scalability, flexibility, the quantity of flow, fairness and network density are the main requirements of a MAC protocol.

10.9.4.1 Energy Efficiency

Energy efficiency is a critical issue because it is highly challenging to recharge or replace the batteries for the many battery-powered nodes.

10.9.4.2 Accommodate Dynamic Nodes

The protocols employed must be able to handle changes in node numbers dynamically because networks may have a large number of nodes. In actuality, their operating programmers must be altered when new positions are created [23].

10.9.4.3 The Quantity of Data Flow

This refers to the volume of data sent in a particular amount of time from a transmitter to a receiver. This rate is influenced by a variety of variables, including latency, channel usage and the efficiency of collision avoidance. The amount of flow is application-dependent, just as latency. Body-sensor network applications frequently call for a long life that accommodates low throughput and high latency [23].

10.9.4.4 Fairness

It is the capacity of various devices or nodes to equitably share the medium. It is a crucial network characteristic since every user needs an equal chance to receive and send data for individual applications. But in WBAN, every node works together to complete a single task. A node might have additional data to share with the other nodes. Instead of treating each device equally, success is determined by the performance of the entire application [23].

10.9.4.5 Network Density

Node densities vary greatly between various applications. Due to the nodes' inactivity or sleep mode, the density can change over time and space even within a single application. Similar to this, the network's density is not uniform, thus it must adjust to these fluctuations [24].

10.9.5 WBAN MAC Protocols

For WBAN, numerous researchers have put out numerous MAC protocols [25]. The IEEE 802.15.4 MAC, Body MAC, Priority Guaranteed MAC, CoCLF-MAC, Battery-aware TDMA MAC, a power-efficient MAC, ECTP-MAC Protocol [26] for WBAN, etc. are a few of these. The benefits and drawbacks of several of these well-known MAC protocols suggested for WBAN are discussed in this section. The energy usage of the MAC protocols is highlighted, along with how they address energy inefficiency brought on by overhearing, collision, control packet overhead and idle listening [27].

10.9.5.1   IEEE 802.15.4 Protocol

To design a standard for WPAN, in September 2006, the Institute of Electrical and Electronic Engineers (IEEE) approved the creation of an IEEE 802.15.4 working organization. These protocols (MAC) are primarily made for wireless low-data transmission applications that operate on the 915 MHz, 2.4 GHz and 68 MHz frequency bands. The PHYS and MAC layers are both defined by 802.15.4. Guaranteed Time Slot (GTS), Channel access, Beacons, management, frame validation, etc., are characteristics of MAC layer management [28]. The MAC layer can operate in either non-beacon mode or beacon mode, in relation to the topology being utilized and the requirement for assured bandwidth [29].

10.9.5.2   IEEE 802.15.6 Protocol

A standard for WBAN which function inside, around or outside the human body is IEEE 802.15.6. It seems to be focused on short-distance, low-cost, dependable wireless data transmission and particularly an extremely low power at relatively low frequencies. The 802.15.6 MAC protocol consists of the SP. The SP has a fixed length and is surrounded by equal-length network beacon periods. The SP arrangement of the IEEE 802.15.6 [26] consists of three operational modes, namely SP bounds in beacon mode with beacon period [29], SP (with boundaries) without beacon mode, and SP (without boundaries) without beacon mode. We attempt to convey the MAC protocols outlined in the most recent draught of the standard in this section.

10.9.5.3   Priority-Guaranteed MAC Protocol (PG-MAC)

The SP organization is crucial in the development of MAC protocols. *PG-MAC* protocol introduces a novel SP organization. Active and idle periods make up the two primary sections of the time axis. To allow various types of information transmission, the Active Section (AC) of the SP is further separated into five sections. Control-Channel-AC1 and Control-Channel-AC2 [26] are used for uplink control of consumer electronics applications and life-critical medical applications, respectively. Time Slot Reserved for Periodic Traffic (TSRP) and Time Slot Reserved for Burst Traffic (TSRB), respectively, are two separate timeslots set aside for a period and burst data. On nodes, synchronization is accomplished using a beacon. For uplink control, AC1 and AC2 use randomized ALOHA. However, TDMA protocol is used to allocate GTS to devices for data connection in the two communication channels. In terms of energy usage, this method performs better as compared to IEEE 802.15.4. The two biggest drawbacks are the difficulty of the SP and the inability to adjust to emergency data flow [30].

10.9.5.4   ELDC MAC

A TDMA-based protocol called Energy-Efficient Low Duty Cycle (ELDC) has been developed to support enormous amounts of data being streamed at once. The lifetime of the network is increased for medium access to optimize the use of the TDMA technique. In the ELDC network topology, the master node is in charge of on-body WBAN synchronization and coordination. There are numerous time slots on the time axis. The criteria employed to determine the amount of reserved communication channels for on-demand application traffic are the packet error rate, the quantity of sensor nodes and the acceptable packet drop rate. To prevent data transmission collisions brought on by clock drifts, guard band slots are placed between two successive slots. In terms of the energy economy, faster data rate and support for brief data bursts, ELDC performs better. Period synchronization, however, will result in increased energy utilization [30].

10.9.5.5 A Power-Efficient MAC (PE-MAC)

The typical, both emergency and on-demand traffic in WBAN can all be accommodated by the PE-MAC protocol. To enhance network performance for both regular traffic as well as on-demand and emergency data traffic, two wakeup techniques are added. In this, the data traffic produced by regular physiological monitoring is classified as normal traffic. Some of the in/on or surrounding the human body sensor nodes start emergency traffic in life-critical applications. The CD asks end nodes to send on-demand traffic so that it can obtain information if necessary. The slotted ALOHA is used during the contention access period to handle a bunch of data. The TDMA is used during the contention-free period to provide collision-free data transmission and make up the time axis in a SP, which is designed to support all three of these communication patterns. Low-power listening, however, is not the best option for increasing the effectiveness of implanted or on-body sensor nodes [31].

10.9.5.6 EMAP

For WBAN to be as energy-efficient as possible, a well-known protocol called Energy-Efficient Medium Access Protocol (EMAP) was created. Periodic scheduling of wake-up and sleep times is done by a central control mechanism. The power loss is minimized via cross-layer optimization. Eight implanted/on-body nodes are regarded to coordinate with star topology network with a single CD. Most tasks and operations are carried out by master nodes. The simulation findings of several physiological indications show that energy consumption is influenced by the quantity of retransmissions of data and the length of sleep intervals. Overhearing and idle listening under central supervision effectively cut down on energy use. However, there are certain implementation restrictions, such as complexity, a cluster's maximum number of nodes, the absence of a mechanism for real-time data and the connection establishment processes [32].

10.9.5.7 PMAC

A frame is a predetermined time that is defined by the PMAC (Proper MAC) protocol. In PMAC, nodes turn on their radios at the beginning of each frame and briefly tune in to the channel. Nodes turn off their radios and go to sleep until the start of the following frame if no packets come within that period. The first is the PMAC proposal for wireless sensor networks, which uses a straightforward data transmission protocol followed by acknowledgment packets sent by the receiver. Without using any additional control signalling of any kind, this is accomplished consistently. By completely eliminating idle listening, the second contribution effectively eliminates the anticipatory characteristic of the MAC layer. The effects of this on energy usage are significant. Its key benefit is its high throughput, but one of its main drawbacks is that it may take a while to adjust to changes [33].

10.9.5.8   Body MAC

To increase power efficiency, Body MAC defines uplink and downlink subframes using the TDMA channel access method. When end nodes cannot communicate any data, they adopt periodic sleep scheduling. The three methods utilized to accommodate various data streams are the periodic bandwidth procedure, burst bandwidth procedure and adjust bandwidth technique. With the help of this adaptable and effective bandwidth management, networks may operate more steadily and transmit control packets more effectively. According to the simulation findings, the Body MAC performs better as compared to IEEE 802.15.4 in considering end-to-end delay and energy conservation as performance metrics. Additionally, it can save energy and provide flexible bandwidth allotment [34].

10.9.5.9   ECTP-MAC

According to the priority of the data flow, F. R. Yazdi et al. presented a MAC protocol to optimize energy usage, collision and latency. ECTP-MAC is the name of this MAC protocol. The structure of this protocol is divided into four sections: data traffic is initially divided into regular, urgent and periodic categories before being prioritized in accordance with those categories. Second, the SP structure of IEEE 802.15.4 has been enhanced, and data priorities have been adjusted. Thirdly, the latency and energy have been optimized by using the radio wake-up technique and controlling the node's modes. Four, the limited ability of the buffers to validate the node's modes has been modelled using the asymmetric Markov model and the state diagram. Regular data are transferred via the CSMS/CA method [30]. The CD additionally sends emergency data using a TDMA phase of the 802.15.4 SP. Finally, regular data transfers occur normally and without any interruptions. The proposed work is simulated by the authors using *ns-3* simulator. According to the simulation findings, ECTP-MAC protocol has lower energy consumption, a lower latency and a longer network life than the 802.15.4 standard [35].

10.9.5.10   CoCLF-MAC

The Fuzzy Logic Control (FLC) for Cooperative Cross-Layer Fuzzy MAC (CoCLF-MAC) in the WBAN context is designed using this MAC protocol technique. This work proposes an FLC (two-level) technique at both the sensors level and the CD level in an effort to further enhance the communication quality in WBAN. The sensors-level FLC controls access based on contention, while the CD-level FLC controls access based on contention-free. This CoCLF-MAC uses a control method (two-level) for WBAN to optimize the dependability of the 802.15.4 structure. Based on the NB rate and required data rate, sensor nodes employ FLC (sensor-level) to adaptively calculate the exponent backoff to avoid a collision. The second FLC has intended to prioritize GTS allocation at the CD level in an adaptive way based on various application requirements and the channel's state. The NBrate and data rate parameters are used to create the first FLC. Data rate and collision rate are used to design the second FLC. This proposed FLC design has the potential to significantly improve the performance and reliability of the network by effectively boosting the coordination between the CD and the sensors. The suggested FLC for WBAN was simulated by the authors using the OMNET++ simulator. The authors evaluated the efficiency and dependability of the proposed FLC using performance indicators such as collision rate, Packet Delivery Ratio (PDR), packet latency and throughput. The simulation findings demonstrate that, in terms of the IEEE 802.15.4 standard, this strategy clearly beats the aforementioned criteria [14].

10.10   CONCLUSION

The current research in WBAN is reviewed in this survey in terms of system design, layer architecture, characteristics, communication technologies, system-level issues and MAC layer for WBAN based on the IoT. A list of currently used sensors, radio technologies, performance indicators and open WBAN concerns is also provided. Several intriguing MAC techniques have been put out in the aforementioned reviews to address the particular characteristics of WBAN, including the use of lengthy propagation delays, signal-based reservations and scheduling-based MAC protocols. Due to the significant challenges and expensive cost of WBAN field trials, none of them has been thoroughly evaluated in practice. The standardization of WBAN MAC protocols and cross-layer design framework is a significant topic that requires additional research. They are useful to focus research efforts on specific WBAN MAC protocols and necessary to ensure interoperability between devices offered by various manufacturers.

REFERENCES

1. S. Kumar, A. Verma, P. K. Verma, and R. Verma, "A Comparative Study of Energy Efficient Collision Avoidance Protocols for WBAN." *2022 IEEE 6th Conference on Information and Communication Technology (CICT)*, Gwalior, India, 2022, pp. 1–5, doi: 10.1109/CICT56698.2022.9997875.
2. V. C. Ferreira, C. Albuquerque, and D. C. Muchaluat-Saade, "Channel-Aware Gait-Cycle-Based Transmission in Wireless Body Area Networks." *IEEE Sensors Journal*, 22(10), 2022, pp. 10009–10017, doi: 10.1109/JSEN.2022.3166695.
3. S. Sharma, and S. K. Awasthi, "Introduction to Intelligent Transportation System: Overview, Classification Based on Physical Architecture, and Challenges." International Journal of Sensor Networks, 38(4), 2022, pp. 215–240, doi: 10.1504/IJSNET.2022.122593.
4. P. K. Verma, R. Verma, M. M. Alrayes, A. Prakash, R. Tripathi, and K. Naik, "A Novel Energy Efficient and Scalable Hybrid-MAC Protocol for Massive M2M Networks." *Cluster Computing*, Springer, 2018, pp. 1–22, doi: 10.1007/s10586-018-1948-y.
5. K. Subramanian, U. Ghosh, S. Ramaswamy, W. S. Alnumay, and P. K. Sharma, "PrEEMAC: Priority Based Energy Efficient MAC Protocol for Wireless Body Sensor Networks." *Sustainable Computing: Informatics and Systems*, Elsevier, 2021, pp. 1–9,doi: 10.1016/j.suscom.2021.100510.

6. P. K. Verma, R. Verma, A. Prakash, and R. Tripathi, "Throughput Enhancement of a Novel Hybrid-MAC Protocol for M2M Networks." *International Journal of Big Data Intelligence*, Inderscience Enterprises, 4(3), 2017, pp. 149–159.

7. M. A. A. Mamun, and M. R. Yuce, "Sensors and Systems for Wearable Environmental Monitoring Toward IoT-Enabled Applications: A Review." *IEEE Sensors Journal*, 19(18), 2019, pp. 7771–7788, doi: 10.1109/JSEN.2019.2919352.

8. P. K. Verma, R. Verma, A. Prakash, and R. Tripathi, "Throughput-Delay Evaluation of a Hybrid-MAC Protocol for M2M Communications." *International Journal of Mobile Computing and Multimedia Communications*, 7(1), 2016, pp. 41–60, doi: 10.4018/ijmcmc.2016010104.

9. P. K. Verma, R. Verma, A. Parkash, A. Agrawal, K. Naik, R. Tripathi, M. Alsabaan, T. Khalifa, T. Abdelkader, and A. Abogharaf, "Machine-to-Machine (M2M) Communication: A Survey." *Journal of Network and Computer Applications*, Elsevier, 2016, pp. 83–105, doi: 10.1016/j.jnca.2016.02.016.

10. X. Cao, Z. Song, B. Yang, M. A. Elmossallamy, L. Qian, and Z. Han, "A Distributed Ambient Backscatter MAC Protocol for Internet-of-Things Networks." *IEEE Internet of Things Journal*, 7(2), 2020, pp. 1488–1501, doi: 10.1109/JIOT.2019.2955909.

11. "IEEE Standard for Low-Rate Wireless Networks," in *IEEE Std 802.15.4-2020 (Revision of IEEE Std 802.15.4-2015)*, pp. 1–800, 23 July 2020, doi: 10.1109/IEEESTD.2020.9144691

12. P. K. Verma, R. Tripathi, and K. Naik, "A Robust Hybrid-Mac Protocol for M2M Communications." *2014 International Conference on Computer and Communication Technology (ICCCT)*, 2014, pp. 267–271.

13. M. Ambigavathi, and D. Sridharan, "A Review of Channel Access Techniques in Wireless Body Area Network." *2017 Second International Conference on Recent Trends and Challenges in Computational Models (ICRTCCM)*, Tindivanam, India, 2017, pp. 106–110, doi: 10.1109/ICRTCCM.2017.28.

14. M. Collotta, R. Ferrero, and M. Rebaudengo, "A Fuzzy Approach for Reducing Power Consumption in Wireless Sensor Networks: A Testbed with IEEE 802.15.4 and WirelessHART." *IEEE Access*, 7, 2019, pp. 64866–46877, doi: 10.1109/ACCESS.2019.2917783.

15. F. R. Yazd, M. Hosseinzadeh, and S. Jabbehdari, "A PriorityBased MAC Protocol for Energy Consumption and Delay Guaranteed in Wireless Body Area Networks." *Wireless Personal Communications*, Springer, 2019, pp. 1677–1695, doi: 10.1007/s11277-019-06490-z.

16. S. M. Nekooei, and G. Chen, "Cooperative Coevolution Design of Multilevel Fuzzy Logic Controllers for Media Access Control in Wireless Body Area Networks." *IEEE Transactions on Emerging Topics in Computational Intelligence*, 4(3), 2020, pp. 336–349, doi: 10.1109/TETCI.2018.2877787.

17. D. D. Olatinwo, A. M. Abu-Mahfouz, and G. P. Hancke, "A Hybrid Multi-Class MAC Protocol for IoT-Enabled WBAN Systems." *IEEE Sensors Journal*, 21(5), 2021, pp. 6761–6774, doi: 10.1109/JSEN.2020.3037788.

18. K. Das, S. Moulik, and C. Y. Chang, "Priority-Based Dedicated Slot Allocation With Dynamic Superframe Structure in IEEE 802.15.6-Based Wireless Body Area Networks." *IEEE Internet of Things Journal*, 9(6), 2022, pp. 4497–4506, doi: 10.1109/JIOT.2021.3104800.

19. A. Nauman, Y. A. Qadri, M. Amjad, Y. B. Zikria, M. K. Afzal, and S. W. Kim, "Multimedia Internet of Things: A Comprehensive Survey." *IEEE Access*, 8, 2020, pp. 8202–8250, doi: 10.1109/ACCESS.2020.2964280.

20. S. Ravidas, A. Lekidis, F. Paci, and N. Zannone, "Access Control in Internet-of-Things: A Survey." *Journal of Network and Computer Applications*, 144, 2019, pp. 79–101.

21. Y. Lu, and L. D. Xu, "Internet of Things (IoT) Cybersecurity Research: A Review of Current Research Topics." *IEEE Internet of Things Journal*, 6(2), 2019, pp. 2103–2115, doi: 10.1109/JIOT.2018.2869847.
22. J. X. Liu, and J. T. Wang, "A MACA-Based Collision Avoidance MAC Protocol for Underwater Acoustic Sensor Networks." *Proceedings of the IEEE/OES China Ocean Acoustics (COA)*, Harbin, China, Jan. 2016, pp. 1–4.
23. Y. Zhang, and G. Dolmans, "A New Priority-Guaranteed MAC Protocol for Emerging Body Area Networks." *2009 Fifth International Conference on Wireless and Mobile Communications*, IEEE, 2009, pp. 140–145.
24. R. Sruthi, "Medium Access Control Protocols for Wireless Body Area Networks: A Survey." *Global Colloquium in Recent Advancement and Effectual Researches in Engineering, Science, and Technology (RAEREST 2016), Procedia Technology*, Elsevier, 25, 2016, pp. 621–628.
25. S. Movassaghi, M. Abolhasan, J. Lipman, D. Smith, and A. Jamalipour, "Wireless Body Area Networks: A Survey." *IEEE Communications Surveys and Tutorials*, 16(3), Third Quarter, 2014, pp. 1658–1686, doi: 10.1109/SURV.2013.121313.00064.
26. S. Movassaghi, A. Majidi, A. Jamalipour, D. Smith, and M. Abolhasan, "Enabling Interference-Aware and Energy-Efficient Coexistence of Multiple Wireless Body Area Networks with Unknown Dynamics." *IEEE Access*, 4, 2016, pp. 2935–2951, doi: 10.1109/ACCESS.2016.2577681.
27. A. Jose, V. A. Tibbie, Pon Symon, and K. R. Shibu, "A Survey on Key Features of Medium Access Control Protocols for Wireless Body Area Networks." *2022 6th International Conference on Trends in Electronics and Informatics (ICOEI)*, Tirunelveli, India, pp. 644–646, doi: 10.1109/ICOEI53556.2022.9776802.
28. S. Saleem, S. Ullah, and H. S. Yoo, "On the Security Issues in Wireless Body Area Networks." *JDCTA*, 3(3), 2009, pp. 178–184.
29. "IEEE p802.15 Working Group for Wireless Personal Area Networks (WPANs): Medwin MAC and Security Proposal Documentation." *IEEE802.15.6 Technical Contribution*, 2009.
30. S. D. T. Kelly, N. K. Suryadevara, and S. C. Mukhopadhyay, "Towards the Implementation of IoT for Environmental Condition Monitoring in Homes." *IEEE Sensors Journal*, 13(10), 2013, pp. 3846–3853.
31. R. Khan, S. U. Khan, R. Zaheer, and S. Khan, "Future Internet: The Internet of Things Architecture, Possible Applications, and Key Challenges." *2012 10th International Conference on Frontiers of Information Technology*, 2012, pp. 257–260, doi: 10.1109/ FIT.2012.53.
32. A. Maatouk, M. Assaad, and A. Ephremides, "Energy Efficient and Throughput Optimal CSMA Scheme." *IEEE/ACM Transactions on Networking*, 27(1), 2019, pp. 316–329, doi: 10.1109/TNET.2019.2891018.
33. T. Wang, Y. Shen, L. Gao, Y. Jiang, T. Ma, and X. Zhu, "Energy Consumption Minimization with Throughput Heterogeneity in Wireless-Powered Body Area Networks." *IEEE Internet of Things Journal*, 8(5), 2021, pp. 3369–3383, doi: 10.1109/ JIOT.2020.3022325.
34. O. Amjad, E. Bedeer, and S. Ikki, "Energy-Efficiency Maximization of Self-Sustained Wireless Body Area Sensor Networks." *IEEE Sensors Letters*, 3(12), 2019, pp. 1–4.
35. D.-R. Chen, and W.-M. Chiu, "Collaborative Link-Aware Protocols for Energy-Efficient and QoS Wireless Body Area Networks Using Integrated Sensors." *IEEE Internet of Things Journal*, 5(1), 2017, pp. 132–149.

11 Internet of Things
Classification, Challenges, and Their Solutions

Sachin Kumar, Pawan Kumar Verma, Rajesh
Verma, Maazen Alsabaan and Tamer Abdelkader

11.1 INTRODUCTION

There are several ways for devices to communicate briefly in today's connected world, including Bluetooth, WiFi, ZigBee, GSM and others. The goal is now to be aware of a variety of actual, non-communicable items in the area as well as to link with other communication devices when the opportunity arises. We may connect different items using the IoT to control, gather, share and analyse the data on their own for use in real-time applications. The Internet of Things (IoT) aims to provide total services to anything, everywhere and at any time via the Internet. IoT plays an important part in most real-time applications, which present the fourth edition of disorderly technology behind the Internet. Smart homes, smart mobility systems, smart health care, industrial automation, smart grids and smart city are the most popular examples of IoT applications [1].

M2M communication is one of the subsets of IoT for connecting things. M2M communication allows direct connection between devices such as mobiles, machines, laptops, etc. by using wired/wireless media. Table 11.1 describes the comparative analysis of IoT and M2M technologies [2]. Using IoT, several things communicate with one another utilizing various technologies and standards such as Wi-Fi, Bluetooth, Wi-MAX and Radio-Frequency-Identification (RFID) [3].

The US National Intelligence Council (NIC) announced that IoT may occupy everyday real-life applications by 2025 such as automatic power control in houses, smart metering, documents, the health sector, etc. The Research and Development Community (RDC) has projected that the impact of IoT on society will be larger than that of the Internet and Information and Communication Technology (ICT), which promotes social and industrial well-being.

Several IoT research problems must be overcome to create a smarter world, maximize IoT efficiency and satisfy application needs. Some of the challenges are energy consumption, scalability, reliability, mobility, interoperability, Quality-of-Service (QoS), system design, data usability and security which affect the potential use of the IoT system. To fulfil the need for IoT, and to optimize the above challenges, a

DOI: 10.1201/9781003452645-11

TABLE 11.1
Comparisons of IoT and M2M technologies

Communication Type			Features			
	Intelligence	Connection Type Used	Internet Connection Used	Accessibility	Common MAC Layer Protocols	Dedicated MAC Layer Protocols
IoT	Devices can make intelligent decisions.	Broadcast	Active internet connection required.	Multiple user connectivity supports.	Pure ALOHA [1], Slotted ALOHA [1], CSMA [8],	ECC [90] Scalable HYB-MAC [8], Hybrid Slotted-CSMA/CA-TDMA [73], etc.
M2M	Limited intelligent decision.	Point-to-point	Devices don't always need to be connected to the internet.	Normally single-user connectivity at a time.	CSMA/CA [8], Scalable HYB-MAC [8].	DPCF-M [8], Adaptive Multichannel Protocol for Large-Scale M2M [8], Adaptive Traffic Load Slotted MACA [8], CERA [8], FASA [8], etc.

common communication infrastructure with well-defined protocols and integration of objects at a large scale is required.

In IoT, many things connect with little to no human involvement. In IoT, MAC protocols play an important role to optimize the above challenges. MCA protocols determine who will acquire the channel when a large number of devices try to access the channel. The main issue in MAC protocol is the collision during channel access. Channel access collision will occur when more than one device tries to access the channel simultaneously. Therefore, an effective MAC protocol is required to address these issues in the IoT and its subset, such as M2M communication.

There have been a number of surveys published that cover different IoT technology aspects. For example, in [4], the authors describe the survey based on the origin of the IoT, challenges in IoT, research topics in IoT and bibliometric studies of the last 20 years. The authors also show the mainstream research area in the field of IoT management, IoT security, IoT challenges and IoT privacy. In [5], the authors provide the main enabling protocols, technologies, application issues, a horizontal overview of IoT and the relationship between IoT and other technologies such as data analytics and cloud and fog computing. In [6], the authors talk about IoT's current research area, IoT characteristics, IoT applications, challenges, issues in the IoT communication standards and future scope in IoT communication. The authors also describe some future research directions including reliability, adaption, context awareness, scalability, interoperability, privacy and security and embedded intelligence. In [7], the authors talk about the comprehensive study of the localization data sources, IoT localization system, localization error, localization algorithms, IoT technologies, localization applications, system architecture and signal measurements. The authors also talk about device diversity, delay, data loss, multipath effects, channel diversity, large area effects, multi-floor effects, base station-based errors and database errors. In [8], the authors describe a comprehensive study of MAC layer requirements, issues and challenges for Machine-to-Machine (M2M) communication. The authors try to describe the existing MAC requirements in terms of scalability, cost-effectiveness, latency, data throughput, energy efficiency and MAC layer protocols for M2M communication. Table 11.2 describes the taxonomy of the last few years' survey papers.

In contrast to prior research, this chapter provides the following contribution:

- To provide a comprehensive study of IoT architecture and the existing as well as some potential applications of IoT.
- To present a comprehensive study of IoT key features, IoT technologies and standards related to recent literature.
- To provide a comprehensive study of the IoT research challenges based on MAC layer protocols and their proposed state-of-the-art solutions.

The remaining sections of this study are structured as follows: Section 11.2 describes the four-layer IoT network architecture. Section 11.3 provides distinctive IoT applications. Section 11.4 provides the key features of IoT. Section 11.5 describes the IoT Technologies and standards. Section 11.6 describes the challenges in IoT and their proposed state-of-the-art solutions. Finally, Section 11.7 concludes the chapter.

TABLE 11.2
Taxonomy of IoT Survey Paper

Survey Paper Title, Year and References	Applications	Characteristics	Protocols	Communication Standards	IoT Challenge
Internet of Things: A Survey on Enabling Technologies, Protocols, and Applications, 2015 [5]	Smart House, Smart City, Smart Industries, Smart Agriculture	X	CoAP, MQTT, XMPP, AMQP, mDNS, DNS-SD, RPL, 6LowPAN, IEEE 802.15.4	X	Availability, Reliability, Scalability, Mobility, Management, Security and Privacy
A Survey of MAC Layer Issues and Protocols for Machine-to-Machine Communications, 2015 [6]	X	X	ALOHA, CSMA, HYBPS, DPCF-M, Scalable HYB-MAC, CERA, FASA	X		Data Throughput, Scalability, Energy Efficiency, Latency, Cost Effectiveness, Coexistence
Machine-to-Machine (M2M) Communications: A Survey, 2016 [2]	Tracking & Tracing, Surveillance & Security, Environmental Monitoring, Automotives, Vehicular Telematics, E-Healthcare, Smart grid	Concurrent Bulk Device Transmission, Bursty Traffic, mobility, Low Power Consumption, Reliability, Priority Scheduling & Access	x	6LoWPAN, 3GPP, ETSI, oneM2M, CoAP, 6TSCH, WiMAX, Wi-Fi, ZigBee, IEEE 802.15.4e	Energy Efficiency, Reliability, Security, Ultra-Scalable Connectivity, Spectrum Management, Quality-of-Service (QoS), Intermittent Connectivity

(Continued)

TABLE 11.2 (CONTINUED)
Taxonomy of IoT Survey Paper

Survey Paper Title, Year and References	Applications	Characteristics	Protocols	Communication Standards	IoT Challenge
Access Control in Internet-of-Things: A Survey, 2019 [98]	Healthcare, Road Management, Surveillance, Industrial Applications	X	Physical and MAC Layer, Routing Protocols	ZigBee, Z-wave, Wi-Fi, BLE, LoRaWAN	X
Internet of Things (IoT) Cyber Security Research: A review of Current Research Topics, 2019, [99]	Smart House, Smart City, Smart Industries, Smart Agriculture	x	HTTP, CoAP, MQTT	RFID, BLE, ZigBee	Security
Multimedia Internet of Things: A Comprehensive Survey, 2020 [100]	Smart Agriculture, Smart City, Home Automation, Healthcare, Industrial IoT, Emergency Care	X	CoAP, MQTT	X	X
Security, Privacy, and Trust for Smart Mobile-Internet of Things (M-IoT): A Survey, 2020 [101]	Smart City, Smart Home, Smart Grid, Personal Care, Emergencies, Smart Factory, Healthcare	Reliable Communication, Ease of Accessibility, High Reachability, Low Power- High Range Transmission, Real-Time Support, Ultra-Dense Connection	x	ZigBee, 6LoRaPAN, Wi-Fi, NB-IoT, BLE, LoRaWAN, 3G, 5G, RPMA, NB-IoT, GSM, WiMAX	The complexity of Design, Interaction Policies, Security, Privacy, Trust, Low-Complexity Protocols, Lifetime

(Continued)

TABLE 11.2 (CONTINUED)
Taxonomy of IoT Survey Paper

Survey Paper Title, Year and References	Applications	Characteristics	Protocols	Communication Standards	IoT Challenge
An Empirical Study on System Level Aspects of Internet of Things (IoT), 2020 [6]	Smart Home, Smart Health, Smart Farming, Intelligent Transportation, Factory Automation, Smart Grid, Factory Automation & Industry 4.0, Environmental Monitoring, Infrastructure Monitoring, Smart City	Internet Connectivity, Network Size, Analysis, Delay, Data Type, Technology Uses, Privacy And Security, Sensors Used, Location Awareness & Sharing	MQTT, CoAP, XMPP, DDS, REST APIs/ HTTP.	RFID, Zigbee, Z-wave, Wi-Fi, NB-IoT, BLE, LoRaWAN, 5G, Light-Fidelity (LiFi), NB-FI, SIGFOX, INGENU, TELENSA	Interoperab-ility, Management, Quality of Service (QoS), Security, and Privacy

11.2   IOT ARCHITECTURE

A basic model of IoT architecture consists of three layers as shown in Figure 11.1 (a). As the daily need for IoT requirements increases, IoT architecture must fulfil the device requirement. So there is a need for a common architecture of heterogeneous IoT devices [5]. In this section, we describe the four-layer common architecture of IoT briefly.

11.2.1   INFRASTRUCTURE LAYER/ SENSOR LAYER

The infrastructure Layer/ Sensor Layer is also called the preparation layer. This layer consists of sensors or devices to collect, process and transfer the data for different IoT devices and enable communication between IoT devices. Devices used for IoT must be reliable, scalable, bandwidth-efficient and energy-efficient to collect information. Devices need connectivity through different gateways with Local Area Networks (LAN), Wi-Fi, Ethernet, etc. [5] [9]. This layer is also called the physical layer in the OSI model, which connects the different devices in the IoT infrastructure [10].

11.2.2   NETWORK/COMMUNICATION LAYER

The network layer also called the transmission layer or communication layer, transfers the data from different objects to the processing unit of the IoT system through a wired or wireless medium [11]. The layer consists of heterogeneous networks (communication and integrated networks) and smartly processes large data [12]. This layer communicates the information by using different technologies such as RFID, Wi-Fi, GSM, Bluetooth, ZigBee, etc. [5] and different protocols such as IPv4/IPv6, etc. [10].

Application layer	Business layer		Protocols
	Application layer		DDS, CoAP, AMQP, MQTT-SN, XMPP, HTTP, REST
Network layer	Network/ communication layer		IEEE 802.11, IEEE802.15.4, 6LoWPAN, IPv4/OPv6
Preparation layer	Infrastructure layer/ Sensor layer	Data Link/MAC layer	ALOHA, SLOTTED ALOHA, CSMA, CSMA/CD, CSMA/CA, PCA
		Physical layer	IEEE802.15.4, BLE, LTE-A, Z-Wave
(a)	(b)		(c)

FIGURE 11.1   IoT Architecture. (a) Three-layer basic model. (b) Four-layer common model. (c) Protocols stack

11.2.3 APPLICATION LAYER

The application layer manages the different IoT applications based on the device information [11] and provides the necessary services and information required by IoT devices [10]. This layer must have the capability to provide intelligent services to heterogeneous IoT devices and applications such as smart farming, smart home, smart city, smart healthcare, smart transportation, etc. [5]. The main challenge at this layer is the safety of the data [12]. Attackers can manipulate the data by targeting the software or algorithms to create misbehaviours [13].

11.2.4 BUSINESS LAYER

This layer is responsible to manage overall application and end-users demands. Based on the information received, this layer provides necessary models, flowcharts, graphs, etc. [11]. The layer is also responsible to develop, monitor, analyse, implement and make smart IoT systems elements. This layer monitors, manages and analyses the output of each layer of IoT systems to provide good quality service [5].

Remarks: IoT is an emerging technology that includes massive numbers of heterogeneous devices with different requirements and different quality of services. Hence the above-explained architecture does not fulfil all the requirements of IoT systems and related technologies.

11.3 DISTINCTIVE IOT APPLICATIONS

The need for the IoT is increasing in day-to-day life applications, hence in this subsection, we elaborate on the popular emerging IoT applications such as homes, industries, farming, health, transportation, automation and industry, infrastructure, cities, environmental monitoring, power grids, etc. [14]. Figure 11.2 shows the most popular IoT application. As the demands of IoT are increasing day by day, Figure 11.3 shows the estimation of the number of connected devices from 2015 to 2025.

11.3.1 SMART HOMES

Automation of homes is a necessity for daily life requirements. Home automation takes home security, electricity meter automation, water-saving, convenience, air quality control, smart gardening, home appliances control such as washing machines, air conditioners, electricity control, etc. IoT needs to control all appliances of the home to the next level [15]. Indoor quality monitoring and control through IoT is essential to improve the oxygen level inside the office or house because indoor air has a higher level of NO, CO_2 and NO_2 [16]. In our busy lives, we can monitor the condition of indoor plants and also control the water supply and nutrition supply through IoT remotely. This is called smart gardening [17]. Home security is a critical issue in our day-to-day life. IoT can provide security and safety through sensors, cameras, alarm systems, motion detectors, real-time video analysis, etc. [18].

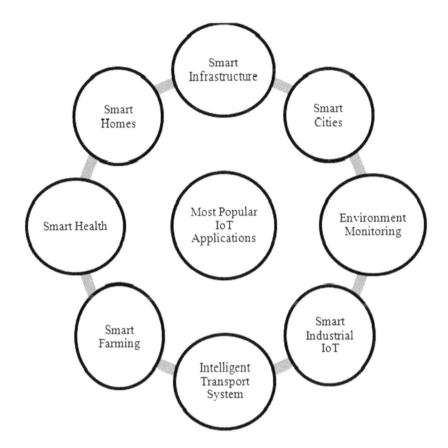

FIGURE 11.2   Most popular IoT applications

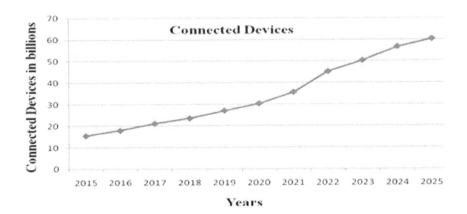

FIGURE 11.3   Estimation of IoT-based connected devices

11.3.2 SMART HEALTH

IoT can provide health-related data smartly and securely anywhere in the world, health remote monitoring, patient care at home, controlling health-related types of equipment remotely, chronic disease treatment and elderly people's health monitoring by using various sensors or devices to collect human body information. Smart health can be used to provide instantaneous medical facilities such as ambulance services to improve health-related facilities [19]. Smart health can improve health facilities through immediate diagnostics, health monitoring in real-time remotely and emergency care by using video conferencing. Smart health played a very crucial role in the COVID-19 pandemic because most doctors were giving online suggestions or treatment to patients.

11.3.3 SMART FARMING

In India, farmers are still using the old types of equipment and methods which cause water wastage, fertilizer waste, soil erosion, etc. [20]. Through IoT, we can control and monitor the condition of crops and also overcome the above issues in farming. The use of sensors and softwares in smart farming can help to control and monitor the climate, soil, water wastage, etc. by analysing real-time information transmitted from IoT devices and controlling the resources for better farming [21]. Smart farming can also be used to monitor the health of the crop (bacterial and fungal diseases) from time to time through IoT-based devices to improve the quantity and quality of the crops. Several IoT-based robots and drones have been developed to monitor the crop's condition and diseases by considering environmental factors.

11.3.4 INTELLIGENT TRANSPORT SYSTEM (ITS)

Smart transportation is one of the biggest applications where we can utilize the maximum benefits of IoT systems such as railways, airports, traffic control management, transport industries, cars, etc. through automation. Smart transportation can save time and money and improve the quality of services, etc. in day-to-day life. Traffic control and monitoring are one of the biggest challenges due to the massive numbers of vehicles in dense cities [22]. Traveller information systems provide information regarding traffic rules, safety, routes, emergency services and accident warnings [23]. Through IoT by using sensors and cameras, we can collect real-time data of transportation systems such as traveller information systems, freight management systems, vehicle-to-vehicle communication systems, etc. In IoT-based ITS, by sharing this information we can overcome transportation problems [22].

11.3.4 SMART INDUSTRIAL IoT

Smart industrial IoT is using real-time data, automation technology, machine learning algorithms, interconnectivity technology and augmented reality to improve overall manufacturing efficiency [24]. Smart industrial IoT also called Industry 4.0

provides services such as reduction in the development process, resource efficiency improvement, production enhancement, automation of resources, plant safety, quality, supply management, etc. [25]. To prevent workers' injuries and dangerous circumstances in plants, surveillance is necessary for plants [24]. Surveillance systems based on IoT can be used to find, analyse and overcome dangerous situations to save human life and the environment. Smart industrial IoT shortens the life cycle of product development and improves the management of the supply chain, management, quality control, etc. [26].

11.3.5 ENVIRONMENT MONITORING

Pollution monitoring in urban areas is required due to the higher population in urban areas. Environment parameters such as humidity, pressure, atmospheric pressure, air quality, $CO2$ and CO must be monitored [27, 28]. IoT devices and sensors allow the monitoring and evaluation of the above parameters to control pollution. Forest loss due to fire (approximately 80%) is a critical issue for the environment, so forest fire monitoring is required to overcome the forest condition. IoT plays a crucial role to monitor forest fires by using heat sensors [29]. Most countries are facing water problems due to water wastage, so smart water distribution is required to save water and the environment. A smart water meter can be used for water distribution [30]. Natural disasters such as hurricanes, earthquakes and floods must be monitored to save lives, properties and habitats. IoT technology such as immediate response systems, warning systems, etc. plays a crucial role in disaster monitoring [31].

11.3.6 SMART CITIES

The smart city plays an important role in enhancing the economy, governance, infrastructure, environmental and social environment, and quality of life [32]. IoT improves the electric grids, transportation infrastructure in the city, water grids and electricity by understanding, analysing, controlling and better utilization of the resources [33]. Smart surveillance plays a crucial role to overcome crimes by using IoT devices such as cameras and drones etc. [34]. The smart city provides the services such as energy, water, fuel and waste management through real-time sensors' data analysis [35].

11.3.7 SMART INFRASTRUCTURE

Smart infrastructure provides real-time data collection, analysis and control related to buildings, tunnels, monuments, bridges, railways tracks, gas and oil pipelines, civil structures, historical structures, aircraft safety systems, power grid systems, dams monitoring systems, etc. to save time, money and to avoid risks [36]. Smart oil and gas monitoring, and monitoring civil structures involve detecting, identifying and processing the damage through IoT sensors. Smart oil and gas monitoring includes detection of leakages and cracks in the pipelines autonomously and civil structure monitoring includes displacement, cracks and stress parameters in the structures [36,

37]. IoT plays an important role in military operations such as identifying the land mine bomb during a search operation by using smart sensors.

11.4 KEY FEATURES OF IOT

The most popular key features of IoT include Artificial Intelligence (AI), sensors, connectivity, small device and active management. Figure 11.4 shows a brief review of IoT features.

11.4.1 ARTIFICIAL INTELLIGENCE

IoT makes things intelligent and smart, i.e., IoT enhances the capability of objects to think intelligently and smartly by collecting, processing and analysing the data through smart algorithms and networks. AI IoT means something is as simple and smart as possible to enhance the daily used objects such as Air Conditioners (AC) and refrigerators to detect the weather condition and run accordingly to save power consumption.

11.4.2 SENSORS

The basic needs of the IoT are sensors. Without sensors, IoT misplaces its importance. Sensors are used to collect information from different aspects, and environments, and send this information to the cloud, database and control system to analyse the necessary action based on different applications [5]. Some of the sensors can be actuators, smart and wearable devices, pressure sensors, chemical sensors, temperature sensors, motion detector sensors, smart cameras, etc. [38].

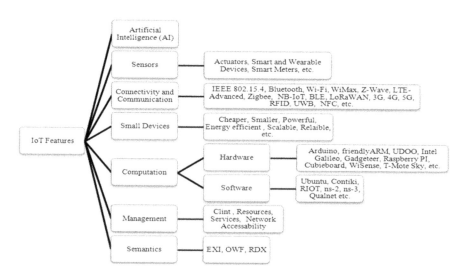

FIGURE 11.4 Features of IoT

11.4.3 Connectivity and Communication

The new enabling IoT technology must connect different heterogeneous devices and different communication standards to provide power-efficient, reliable, secure and intelligent services. IoT uses different communication standards to collect and share information across the world. Some of the communication technologies and standards that can be used for IoT are IEEE 802.15.4, Bluetooth, Wi-Fi, WiMAX, Z-Wave, LTE-Advanced, ZigBee, NB-IoT, BLE, LoRaWAN, 3G, 4G, 5G, RFID, Ultra-Wide Bandwidth (UWB), Near-Field Communication (NFC), etc. Bluetooth, Wi-Fi and WiMAX use radio waves for communication and provide a range of up to 10 m, 100 m and 50 km, respectively [39]. The IEEE 802.15.4 is used to provide energy-efficient, reliable, scalable and secure communication [40]. RFID uses a simple chip for object identification and provides a range of 10 m to 200 m. The NFC technology uses a 13.56 MHz band to provide short-range communication up to 10 cm [41].

11.4.4 Small Devices

Devices used in IoT networks must be cheaper, smaller, more powerful and energy-efficient to deliver scalability, precision and versatility. Some of the devices can be actuators, smart and wearable devices, pressure sensors, chemical sensors, temperature sensors, motion detector sensors, smart cameras, etc. [38].

11.4.5 Computation

IoT networks use microcontrollers, microprocessors and embedded systems to collect and process information and OS such as Ubuntu, Contiki, RIOT, etc. to analyse the data. Arduino, friendly ARM, UDOO, Intel Galileo, Gadgeteer, Raspberry PI, Cubieboard, WiSense, T-Mote Sky, etc. are some of the hardware used for IoT applications [42]. Cloud computing plays an important role in IoT-based computational systems to store their data, analyse the data and extract the big data for real-time applications.

11.4.6 Semantics

The extraction of information intelligently from different objects (machines) in an IoT-based system is called semantics. Semantics includes discovering, information modelling, recognizing, analysing information and making intelligent decisions to provide real-time and accurate services [43].

11.4.7 Management

Management in IoT refers to managing the IoT devices, IoT clients (i.e., managing the numbers of clients), required services, accessibility of the network, etc. IoT device management manages the number of devices, operation of devices and maintenance of devices in the network [44]. Service management includes information aggregation services (i.e., sensing, gathering, storing and analysing the information), collaborative aware services and ubiquitous services management.

11.5 IOT TECHNOLOGIES AND STANDARDS

IoT standardization is required to develop universal IoT systems architecture and to provide energy-efficient, reliable, scalable and cost-effective communication for different applications [2]. Wired and wireless standards are commonly used for IoT communication. Wireless IoT standards can be categorized as short-range communication and long-range communication standards such as IEEE 802.11, IEEE 802.15.4, IEEE 802.15.1, etc. [45]. In this section, we will discuss the major standards developing organizations and also discuss the major standards used in IoT communication.

11.5.1 IoT STANDARDS DEVELOPING ORGANIZATIONS

To provide seamless connectivity in IoT, devices must follow a well-defined protocol and standards. Several Standard Development Organizations (SDOs) build the standard protocols, standards and technologies to provide seamless connectivity for IoT devices. Some examples of SDOs are 3GPP, IEEE, ITU and IETF, used to define and resolve problems in the technology [2].

11.5.1.1 International Telecommunication Union (ITU) Standardization for IoT

SG20 (Study Group 20) defines the IoT applications such as smart cities and smart communities to provide seamless connectivity, identification, privacy and security in IoT systems. ITU also defines reference network architectures to provide common solutions for different IoT applications such as e-health, e-agriculture, smart manufacturing, industrial IoT, etc. ITU can be used to define the spectrum for IoT and its applications.

11.5.1.2 IEEE Standardization for IoT

The IEEE organization is producing standards for wired and wireless connectivity (e.g., LAN, WAN, PAN, etc.) for defining physical and MAC layer protocols to provide a new specification to fulfil the requirement of different users. IEEE 802.11 standard (Wireless LAN) is a MAC layer standard that can be used for many IoT applications. However, IoT-based application demands low-power standards for low-power transmission because most of the devices are battery-operated. IEEE introduced IEEE 802.15.4 standards for low-power operation (LowPAN). IEEE tries to define futuristic standards for physical and MAC layer specifications for interoperability between heterogeneous devices and for short-range communications [46].

11.5.1.3 Third-Generation Partnership Project

The Third-Generation Partnership Project (3GPP) service and System Aspects (SA1) are working on the optimization of networks for IoT and its applications in terms of the overall cost of the systems. The SA2 maintains the network connectivity for the objects and information exchange by the objects. 3GPP-SA3 maintains the security issues of the IoT systems. 3GPP-Release 10 is used for optimization addressing, energy efficiency, congestion, identifiers, subscription control, over-load control and security. 3GPP-Release 11 has been working on the optimization of peer-to-peer communication, IoT gateway, improvement for IoT groups, network navigation, service requirement, etc. [2]. 3GPP also uses narrowband IoT (NB-IoT) to produce a specification for cellular communication. In 2016, 3GPP produced a radio standard for NB-IoT developed for low-power WAN. NB-IoT is used for energy-efficient battery life and low-cost indoor IoT [46]. The 3GPP standards delivered enhanced data rates for global evolution (EDGE) for high-speed packet access coverage (653 uplink and 1.3Mbps downlink). 3GPP2 standards have been used for high-speed, internet protocol and broadband-dependent mobile systems to provide global roaming [2].

11.5.1.4   Internet Engineering Task Force (IETF) Standardization for IoT

The IETF organization aims to produce internet protocols for IoT networks at different levels for monitoring and building smart grid infrastructure [2]. IETF is an IEEE, 3GPP and ITU-affiliated organization that produced enabling protocols to connect the nodes on physical and data link layers [46]. IETF creates different activities related to intelligent objects such as Routing Over Low-power and Lossy networks (ROLL), 6LoWPAN and sensor technologies. ROLL groups focus on a low-power routing protocol for low power, limited memory and limited processing capacity based on embedded device nodes. IEEE 802.15.4, Bluetooth and Wi-Fi technologies are used to interconnect these nodes to provide seamless connectivity between nodes. 6LoWPAN (IPv6 over low-power WAN) aims to provide internet protocol facilities to low-power-based small devices in IoT networks [2].

11.5.2   IoT Technologies Standards

In this subsection, we discuss common standards that can be used for IoT communication such as IEEE 802.15 Group Standards, ZigBee, RFID, Bluetooth, Wi-Fi, WiMAX, LoRaWAN, SIGFOX and NB-IoT. The classification of IoT standards is shown in Figure 11.5.

11.5.2.1   IEEE 802.15 Group Standards

The IEEE 802.15.4 standard was published by IEEE 802.15 group for LR-WPANs (low rate WPAN) physical and MAC layer-based specifications. ZigBee, Wireless HART and Mi-Wi specifications were founded by this group. The IEEE 802.15.4e was developed to fulfil the requirement of industrial applications such as building automation, smart grids and factory automation by modifying the existing IEEE 802.15.4 MAC specification. The main work of the IEEE 802.15.4e group is to optimize link reliability and channel security [2]. IEEE 802.15.4f active RFID system standard was approved in 2012 to provide low-cost, highly reliable and ultra-low energy consumption communication for sensors and active RFID-based applications. The IEEE 802.15.4g standard also called a smart utility network is developed to provide a global standard for process control large-scale applications by modifying the physical layer of IEEE 802.15.4 standards [2].

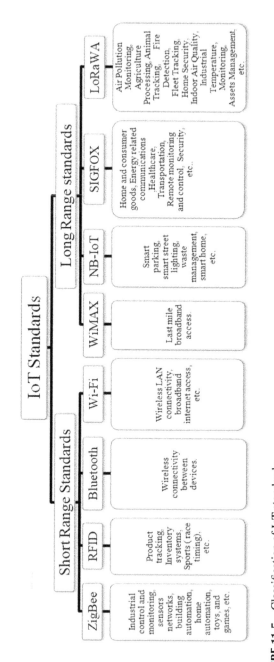

FIGURE 11.5   Classification of IoT standards

11.5.2.2 ZigBee

ZigBee Alliance developed the ZigBee standard to carry low power, low cost and small two-way data for IoT communication. ZigBee technology can provide 150 meters range, 1 MHz bandwidth and 250 kbps speed for smart homes, smart industry and small area-based applications communication [47]. ZigBee standards use IEEE 802.15.4 physical layer standards to specify network and application layer specifications [48]. ZigBee defines functional procedures and message formats to provide interoperability by using different application profiles [49].

11.5.2.3 RFID

RFID is used for tracking, identifying and collecting information from various objects by using wireless communication [50]. RFID uses a combination of microchips (tags) and readers to detect objects. RFID microchips are used to store identification-related information to detect objects. RFID technology can detect objects up to 10 meters range by fetching the information when tags come within the range [50].

11.5.2.4 Bluetooth (IEEE 802.15.1)

Bluetooth is a short-range wireless communication technology that uses RF waves to connect different heterogeneous devices such as mobile phones, laptops, smart homes, offices, Bluetooth-enabled devices, etc. Bluetooth is energy-efficient technology that operates in 2.4 GHz ISM (industrial, scientific and medical) bands and provides high data rates of up to 1Mbps. Bluetooth devices use a frequency range from 2.404 GHz to 2.480 GHz total of 79 1MHz frequencies [51]. Bluetooth technology uses IEEE 802.15.1 standards to define physical and MAC layer specifications [2].

11.5.2.5 Wi-Fi

Wi-Fi technology used 802.11 standards to provide radio connectivity for a short distance (up to 50 meters) by using Wi-Fi-enabled devices such as routers. Wi-Fi technology provides a data rate of up to 11 Mbps [52]. Wi-Fi uses 60 GHz, 5 GHz, 3.6 GHz and 2.4 GHz bands and physical and MAC layer specifications to provide energy-efficient and secure wireless communication [2]. IEEE 802.11 standards involved 802.11a, 802.11b, 802.11g, 802.11i, 802.11n, 802.11ac, 802.11ah and 802.11ax versions that provide different ranges of communication [2].

11.5.2.6 WiMAX

The IEEE 802.16 standard called WiMAX (worldwide interoperability for microwave access) technology was developed to provide Point-to-Multi-Point (PMP) broadband wireless access. WiMAX technology has overcome the gaps between WAN and LAN technologies. WiMAX technology provides low-cost, extended coverage, a large number of users and high bandwidth communication. The IEEE 802.16d standard provides fixed broadband access and IEEE 802.16e technology provides mobile broadband access in the MAN network. WiMAX technology provides a high data rate, good spectral efficiency and a good communication range of up to 50 km [2].

11.5.2.7 LoRaWAN

LoRaWAN is an energy-efficient long-range wireless communication technology used to connect battery-operated IoT devices [53]. LoRaWAN uses three channels (868.10, 868.30 and 868.50 MHz) in ISM bands to provide a low data rate of up to 50 kbps and a communication range of 5 km to 20 km [53, 54].

11.5.2.8 SIGFOX

SIGFOX uses 433 MHz, 868 MHz and 915 MHz frequencies in Asia, Europe and North America, respectively, to provide low-power communication technology for IoT. SIGFOX provides data rates of 100 bps up to a communication range that varies from 10 km to 40 km [55]. SIGFOX uses UNB (ultra narrowband) modulation techniques to improve the receiver sensitivity and overcome energy consumption. SIGFOX technology-based devices can transmit up to 140 messages per day [56]. The SIGFOX MAC layer sends three messages by using a 200 kHz macro channel in the ISM band [57].

11.5.2.9 NB-IoT

3GPP developed a low-power WAN cellular technology called Narrowband IoT (NB-IoT). NB-IoT uses LTE standards and provides low cost, low power and a large area for IoT networks [58]. NB-IoT technology can connect a large number of devices (50,000) and provides 250 kbps uplink speed and 230 kbps downlink speed by using a single-carrier frequency division multiple access (SC-FDMA) schemes. NB-IoT can cover up to a 35-km coverage area [59]. NB-IoT technology uses 180 kHz bandwidth that is divided into 12 subcarriers (15 kHz each). NB-IoT provides two types of transmission called single-tone and multi-tone transmission [54].

11.6 IOT RESEARCH CHALLENGES AND THEIR PROPOSED STATE-OF-THE-ART SOLUTION

The development of the IoT network using a massive number of intelligent devices to fulfil the future requirements of users and IoT applications is a critical task. IoT devices used processing, sensing, actuating and communicating capabilities to provide smart services. To provide seamless services in IoT, issues such as availability, reliability, scalability, mobility, privacy, security and interoperability must be optimized to achieve scalable, reliable, energy-efficient, cost-efficient and ubiquitous communication [2, 60]. The MAC layer protocols for IoT networks must be designed with a rich set of demands to fulfil the above requirements. In this section, we discuss the challenges related to IoT communications [2]. Figure 11.6 shows the IoT research challenges.

11.6.1 AVAILABILITY

Availability in IoT refers to how easily an IoT user can access software and hardware to access services anywhere, at any time. Availability of software means the ability

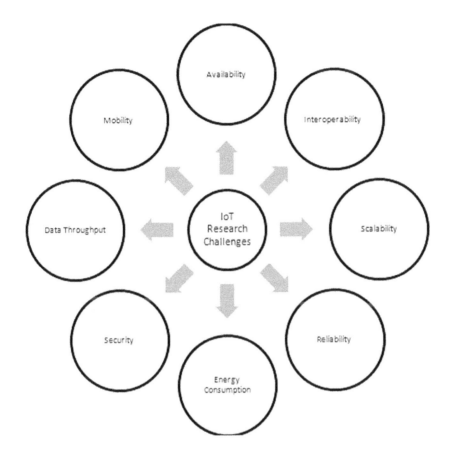

FIGURE 11.6   IoT challenges

of the systems to provide services for massive devices simultaneously anywhere, at any time. Availability of hardware means, easy availability of IoT devices, and how easily these devices are compatible with IoT systems and technology used for required data [61]. IoT hardware must be embedded with IoT-enabled protocols such as IPv6, RPL, 6LoWPAN, CoAP, etc. to provide the IoT network services [5].

To optimize the availability, in 2014, D. Macedo et al. [62] proposed a model to estimate the availability and reliability by providing redundancy to IoT-based applications. In the proposed model, the authors use spare devices to optimize the availability of data during primary device failure. To create such models, a collection of discrete system states is defined and a series of transition rates varying from one state to another is determined. In this technique, the failure rates and repair rates are used to configure the transition rates.

Further, in 2016, Q. Wei et al. [63] produced a method to provide service availability by context capturing. The authors used an environmental-based modelling method by using WSMO-Lite (for dynamic contest modelling and service availability) to provide IoT services. The authors enhanced the WSMO-Lite to enable modelling of the dynamic environment and service availability. First, a context-capturing environmental entity model is provided. Second, a three-dimensional representation of service accessibility is provided: (1) the service model has two extra properties, namely location and available time; (2) a QoS model for resources and mobility is attached; (3) the dynamic service behaviour has been modelled using probabilistic automata.

Further, in 2018, W. B. Qaim et al. [64] proposed a data availability technique for sensors-based IoT networks. This technique known as DRAW (distributed hop-by-hop data replication technique) enhances data availability during node failures. The DRAW technique chooses the finest replica network node based on information provided by neighbours to save a copy of the available data. DRAW also enhances the quality of data distribution by achieving improved replica spread in the network in terms of the distance travelled by subsequent replicas from the source node. This DRAW ensures optimum data availability to protect data in the event of high node failures. The authors use the ns-3 simulator to analyse this technique and show that DRAW enhances the data availability by 15% as compared to the Greedy technique. For critical IoT applications such as military applications, there must be middleware support available all the time to support system failure. The availability and reliability must provide the maximum fault-handling capability for IoT applications [65].

11.6.2 RELIABILITY

Reliability is an important issue in IoT communication systems that ensures the delivery of data success (probability of success) and proper working of the system, which aims to improve the success rate in IoT communication [5]. In IoT systems, reliability must be as high as possible in terms of data collection, processing and transferring data successfully. In a wireless communication system, reliability can be achieved by optimizing channel fluctuations, collision reduction, noise reduction and reduction in end-to-end delay [2]. Reliability is correlated with availability, as it ensures the overall services of the network. In some time-critical applications, unreliable information can harm human life or system failure. In an IoT-based system, the reliability of information is important at each layer of the system model implemented in hardware as well as in the software part of the system [66].

To optimize the reliability, in 2011 M. Zayani et al. [67] studied the IEEE 802.15.4 model for better understanding by presenting IEEE 802.15.4 mimics model at physical and MAC layers. This work used Krishnamachari and Zuniga's mathematical framework and optimized the link quality in terms of distance, asymmetry, noise, output power and errors related to modulation and encoding. The physical model avoids the disk-shaped node range and expresses the level of connection unreliability with greater precision. This model output is significant because it allows for a more accurate estimation of the frame failure probability. The authors also try to improve the IEEE 802.15.4 model reliability and end-to-end delay by improving Park et al. [13] approach and combining this with the physical layer model. To more accurately estimate the wireless network parameters, this work enhances the Park et al. method to figure out frame transmission probability and merge it with the PHY model.

Further, in 2018, M. P. R. S. Kiran et al. [68] developed a new model to examine the IEEE 802.15.4 MAC standard with a size-constrained packet queue. They also proposed an IEEE 802.15.4-based power efficient scheme to increase reliability in terms of power at the node. The authors used the 3D Markov model and M/G/1/K queue to improve the MAC protocol (CSMA/CA) and packet queue (on-node) to evaluate the performance of packet queue overflow losses, precise packet service time, delay, reliability and power consumption of the node. The authors use C programming to simulate the scheme and claim that the proposed scheme achieves 97% accuracy and energy-saving efficiency of up to 74.82%.

Further, in 2018, Y. Jin et al. [69] proposed a content-centric cross-layer routing and scheduling solution (CONCISE) for industrial IoT applications to route and aggregate data by creating schedulers and content-independent routing topologies, resulting in numerous content-based scheduling and a layered routing system. This scheme also handles data aggregation and in-network processing to decrease network traffic. This CONCISE scheme uses content types to divide the network routing and MAC scheduling. In this, each type of content and application has its own independent unique schedules and routing tree, as a result, CONCISE creates several layered x-layer schedules for various content flows. In this CONCISE, the authors use content-specific ways via deterministic Time Synchronized Channel Hopping (TSCH) to reduce the network traffic (65%), end-to-end delay (50%) and to improve the reliability as compared to 6TiSCH scheduling solutions in IoT. The authors use the Contiki Cooja simulator, MSP430F5438 CPU and CC2420 radio based on TI's exp5438 platform to examine the performance of the given solution.

Further, in 2019, A. Bakshi et al. [70] proposed a MAC protocol to improve short-lived demands and high-intensity IoT networks called EMIT (efficient MAC paradigm for IoT). EMIT bypasses the coordination cost and high overhead of MAC protocol by sharing resources simultaneously. EMIT uses the dense and dynamic character of IoT networks to reduce the spatio-temporal variability of interference and achieve minimal delay and high dependability in service. In EMIT, each device transmits the data by selecting the optimal data rate and low power level. Due to limited transmission power, data are encoded slowly over time. The authors use delay and reliability as performance metrics. To examine the algorithm, the authors develop TelosB Testbed and Orbit Lab USRP Testbed and show that EMIT improves the above metrics as compared to the CSMA MAC protocol.

11.6.3 SCALABILITY

Scalability is another challenge for IoT systems because a large number of devices want to comfortably communicate with each other. Scalability in IoT can be defined as the "ability to manage the massive number of devices, adding new devices, services, and functions for customers without affecting the system performance" [6]. Managing a large number of sensors or devices in wireless communication is not an easy take [5]. The tradeoff between scalability and overall system throughput

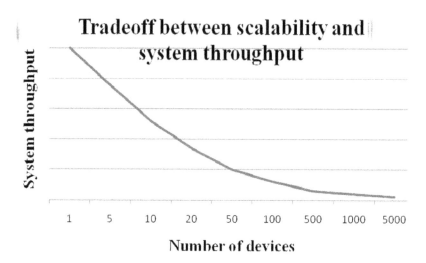

FIGURE 11.7 Tradeoff between scalability and system throughput

is shown in Figure 11.7. In IoT, scalability can be optimized by using cloud, fog and edge computing technology and by improving the waiting time to transmit the data [6].

To optimize the scalability, in 2013, F. Li et al. [71] proposed an IoT-based Platform as a Service (PaaS) framework to optimize the scalability and efficiency of the heterogeneous devices used in IoT architecture. This proposed architecture uses computational resources to provide scalable new services. The PaaS offers cloud platform services for IoT solution providers to effectively provide and constantly improve their services. The implementation and design of this framework support the idea of virtual verticals by inheriting the multi-tenant nature of the cloud. Each user of an IoT solution in virtual verticals has access to a virtually isolated solution that they can tailor to their specific physical settings and devices while sharing the middleware services and underlying computational resources with other users. This work is implemented on the open-source WSO2 Stratos platform.

Further, in 2014, C. Sarkar et al. [72] addressed the fundamentals for a generic IoT architecture and based on this, a distributed layered architecture (DIAT) is being presented for heterogeneous devices connectivity. This architecture provides scalability and interoperability in IoT networks, i.e., to connect a large number of heterogeneous devices. This architecture has a virtual object, service layer and composite virtual objects. The DIAT uses the main functionality of three layers of the architecture of the IoT called the Virtual Object Layer (VOL), *Composite Virtual Objects* (CVO) *layer,* and *Service Layer* (SL). In this scheme, the virtualization of physical objects or entities is the responsibility of the VOL. The CVO layer performs the coordination and communication among multiple objects and the maintenance and creation of services are the responsibility of the SL. This architecture is being tested on a working project icore.

Further, in 2018, N. Shahin et al. [73] proposed a MAC layer protocol for massive IoT device registration called hybrid slotted-CSMA/CA-TDMA (HSCT) protocol for M2M networks. In this, the authors concentrate on scenarios in which numerous M2M devices attempt to register at a single, centralized AP simultaneously. The devices use slotted CSMA/CA protocol to send authentication signals through randomly chosen backoff slots, whereas TDMA protocol is used to send and receive association requests and responses through the allocated slot. The authors use multiple CSMA/CA access windows for network compatibility and to avoid congestion when a large number of M2M devices try to do registration. The authors compare the proposed HSCT protocol with the IEEE 802.11ah standard and Contention-Free Transmission (CFT) in terms of robustness and scalability. The authors use the ns-3.24 simulator to implement IEEE 802.11ah MAC. The simulation results show that the proposed HSCT protocol takes 64% and 87% less time (delay) in massive device registration as compared to the CFT-Adaptive Threshold Algorithm (CFT-ATA) and CFT mechanisms.

11.6.4 INTEROPERABILITY

Interoperability is one more issue in IoT systems due to maintaining good QoS, data transmission formats and different technology, different new IoT applications and underlying communication standards between massive numbers of heterogeneous devices [74]. In a heterogeneous wireless environment, different users demand different requirements so it is a big challenge to design a protocol to meet customers' requirements. Interoperability refers to the ability to interchange and uses data across different hardware and software systems [75]. Interoperability must be managed at every layer of the system. Device interoperability, network interoperability and syntactic interoperability must be managed at the perception layer, transport layer and application layer, respectively. Interoperability must be managed due to heterogeneity in available devices, heterogeneity in different IoT-based applications and different techniques used for communication [75].

To optimize interoperability, in 2017, S. Schmid et al. [76] proposed an architecture based on platform interoperability called cloud-based BIG architecture. The architecture makes use of semantic web technologies to promote interoperability between IoT apps, platforms, devices, products and services and to provide interconnection between different APIs platforms. This architecture provides resource registration and authentication features.

Further, in 2019, S. Yang et al. [77] proposed a new semantic interoperability strategy called the Tabdoc approach to handling a cross-context end user-objects interoperability problem. This approach uses a transaction interface between external and internal objects and users to control information access. To understand the consistent message and to implement this approach, a semantic interpretation algorithm and a semantic extraction algorithm are used. The Tabdoc approach uses the "divide and conquer" tactic to separate a complicated document into three levels: separate models for syntax, instances and templates from one another. The purpose of the Tabdoc technique is to ensure that (1) real devices may receive commands from device users without ambiguity and (2) services provided by actual devices can be delivered without ambiguity to device users. This method specifically uses the

Tabdoc to describe semantic documents with heterogeneous data in a semantically consistent manner across contexts.

11.6.5 ENERGY CONSUMPTION

Energy efficiency (reduction in energy consumption or energy saving) is an important design consideration for IoT networks because most of the nodes in IoT systems are battery-operated [78]. Energy efficiency is a critical issue to maximize network performance. The tradeoff between scalability and energy consumption is shown in Figure 11.8. Reduction in energy consumption or energy saving in IoT networks can be achieved by optimizing the sensing, processing, data transmission, energy distribution between the nodes, reduction in collisions, sleep schedule and reduction in idle listening [79].

To optimize energy consumption, in 2015, H. Park [80] developed a common channel access algorithm called Adaptive Backoff Enabled-Wake-up Radio (AB-WUR) for edge devices that improve energy efficiency (low power consumption) and channel access delay. This scheme used traffic load based on the current network situation in which the contention window value is determined by the access point. AB-WUR scheme sends a message to the WUR device instructing it to execute load balancing by accessing the channel based on the traffic load. The authors mathematically calculate the network parameters such as collision probability ratio, throughput and packet delivery ratio. The authors use discrete event simulations and the results show that AB-WUR throughput is twice as compared to general WUR throughput.

Further, in 2018, M. P. R. S. Kiran et al. [81] proposed a new model to enhance the MAC layer performance of IEEE 802.15.4 standards to give a new MAC layer Prioritized Contention Access (PCA) protocol for time-critical data transmission. The authors use both non-bacon enabled and bacon-enabled MAC protocols of 802.15.4 standard. In this model, the authors developed a Markov chain model for

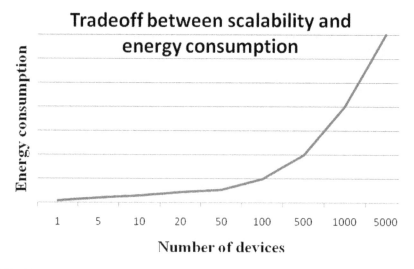

FIGURE 11.8 Tradeoff between scalability and energy consumption

slotted and unslotted CSMA/CA and PCA to improve end-to-end delay and power consumption without any reliability loss. To examine the performance of the given model, the authors developed a testbed using IITH Motes and Contiki 3.0 and the analysis results showed that the slotted and unslotted PCA reduced the delay (63.3% and 53.3%, respectively) and power consumption (97% and 96%, respectively) as compared to slotted and unslotted CSMA/CA.

Further, in 2019, X. Lin et al. [82] presented the data collection issues based on energy consumption for Unmanned Aerial Vehicles (UAVs) IoT called UAV-IoT. The authors tried to improve the energy efficiency of PHY-MAC parameters by reducing the system throughput. The authors tried to make a balance between energy efficiency and system efficiency to optimize the network performance by using the Particle Swarm Optimization (PSO) method to find the optimal value and find that the boundary point is the optimum value point.

Further, in 2020, N. Choudhury et al. [83] described a comparative study of Time-Slotted-Channel Hopping (TSCH) and Deterministic and Synchronous Multi-channel Extension (DSME) MAC models and IEEE 802.15.4 MAC by using performance parameters such as latency, throughput and the cost in terms of energy. The authors also proposed the TSCH CSMA/CA-based Markov model to estimate transmission delay and power consumption. They analysed the performance in terms of energy efficiency, QoS requirements, reliability and end-to-end latency. The authors used MATLAB to implement the IEEE 802.15.4 MAC and TSCH. The analysis results showed that the DEME and TSCH are better in terms of latency and throughput as compared to 802.15.4 MAC. For relaxed throughput and latency, IEEE 802.15.4 is better in terms of energy efficiency than TSCH and DSME.

Further, in 2020, O. Said et al. [84] proposed an energy management scheme (EMS) for IoT networks considering heterogeneous nodes. In this EMS, the authors use three strategies to minimize the data volume to schedule the work of critical energy nodes and to give a scenario for fault tolerance that overcomes energy problems of IoT nodes. To evaluate the performance of the proposed scheme the authors use energy consumption rate, throughput, number of failed nodes and network lifetime as performance metrics and simulate the proposed scheme using ns-2 simulator.

Further, in 2021, B. Safaei et al. [85] proposed an energy-efficient cross-layer objective functions MAC layer protocol called ELITE. This protocol defines a routing metric that shows the number of per-packet transmitted strobes. In this protocol, ELITE chooses a less strobe path for transmission. This protocol reduces the energy consumption for the device node by transmitting fewer strobe packets in the radio duty cycling mechanism of the MAC protocol. To evaluate the protocol, the authors use the Cooja simulator (Contiki's in-build IoT simulator) and found that the proposed ELITE protocol reduces the strobes per packet by up to 25% efficiently and improves the average energy consumption by up to 39% in IoT.

11.6.6 MOBILITY

Mobility is another challenge in IoT-based systems because mobile devices create connectivity problems, handover issues, desired Quality of Services (QoS) and bandwidth issues, and service interruption issues when the device moves from one gateway to another [5]. The IoT has a large number of heterogeneous devices with different issues such as radio technologies interfaces (Wi-Fi, Wi-MAX, Bluetooth, etc.), management of seamless mobility and handover issues are critical challenges [86]. So there is a need to optimize routing protocol to overcome the mobility issues in IoT communication networks.

In this regard, in 2019, K. Cao et al. [87] proposed an energy-efficient network lifetime optimization scheme for battery-operated mobile devices. The suggested approach is divided into two stages: offline and online. To optimize IoT network lifetime, a mobility-aware task plan and mixed-integer linear programming are used in offline mode. To adapt task execution to variable QoS requirements, a cross-entropy method based on a time-efficient, performance-optimized and QoS-adaptive heuristic is presented. The MPSoC systems are used for synthetic and real-life application-based simulation. The simulation analysis results show that the proposed scheme optimizes the network lifetime (169%) as compared to benchmark solutions.

Further, in 2019, A. A. R. Alsaeedy et al. [88] proposed a new scheme to optimize the paging and Tracking Area Update (TAU)-based mobility management in 5G and LTE technology-based wireless networks. For highly mobile IoT devices and users, paging and TAU procedures increase the power consumption and signalling overhead for battery-operated devices. The authors propose gNB-based User Equipment Mobility Tracking (UeMT) solution to optimize the above problems. This solution bypasses the TAU procedure in mobility tracking (tracking of IoT devices will be performed by gNBs) which gives no signalling overhead and power savings. This scheme also improves the paging procedure by providing high reachability with a reduction in the paging message. In this scheme, signalling overhead is optimized by 92%.

Further, in 2020, Di Wu et al. [86] presented ubiquitous flow control (UbiFlow)-based software for mobility management and omnipresent flow control for heterogeneous IoT networks in an urban area. To provide distributed control of IoT flows, UbiFlow uses several controllers to divide urban-scale SDN into distinct geographic segments. The key challenges that depend upon mobility management, such as fault tolerance, load balancing and scalable control, have been rigorously explored and studied based on this UbiFlow overlay structure. Flow scheduling is differentiated by the UbiFlow controller based on per-device requirements and whole-partition capabilities. As a result, multi-network can display a network status view and an optimized selection of Access Points (APs) to satisfy IoT needs while ensuring network performance. The OMNeT++ network simulator and experiments on a realistic testbed are used to analyse the performance. The analysis results show that UbIFlow successfully optimizes the mobility in IoT networks (i.e., 72.99% delay deduction, 67.21% throughput optimization and 69.59% jitter enhancement) as compared to SDN systems.

11.6.7 DATA THROUGHPUT

In IoT systems, another requirement is the high throughput for real-time applications. In MAC layer-based channel access mechanism, an end-to-end delay occurs due to limited channel/spectrum resources availability, channel access mechanism, collision in data transmission, control overhead and empty slots which degrade the system throughput. The throughput optimization is necessary to provide seamless connectivity in IoT-based applications [81, 1].

In this regard, in 2015, D. Striccoli et al. [89] proposed a Markov model to improve the delays, throughput and frame losses in IEEE 802.15.4 MAC layer channel by tracking the channel behaviour at the frame level. The proposed k-state model uses long-range correlation properties and high variability of the channel conditions. The authors analyse the reproduced data traces in a really noisy environment, generate a synthetic data trace and compare the main trace statistics and real traces to calculate the relative errors. The simulation results show that the proposed model improves the Hidden Markov Model (HMM).

Further, in 2016, D. D. Guglielmo et al. [90] proposed an approach to optimize the performance of IEEE 802.15.4 in terms of accuracy and tractability for large-scale networks in Non-beacon-enabled (NBE) mode. This new technique is termed Event Chains Computation (ECC). This technique is used to model the unslotted CSMA/CA protocol in NBE mode. In this scheme, the CSMA/CA algorithm results can be described as a series of events that occur in the network. This ECC protocol can generate all the conceivable outcomes and probabilities that a CSMA/CA algorithm can provide. The ECC considers only those events whose occurring probability is greater than the predefined threshold to overcome the complexity. The authors use the Tagged node model using a single node and Stochastic Automata Network (SAN) model by using all nodes to analyse delivery ratio, delay distribution and average packet latency. This ECC algorithm uses a set of probable outcomes to analyse the delay, delivery ratio and energy consumption as performance metrics and analyses the model through testbed (Tmote-sky motes) experiments and Contiki OS simulations. The authors observe that for low contending nodes, the delivery ratio is very low which can be optimized by increasing the backoff stages and initial backoff window size allowed for each packet. The authors demonstrate that the computation time necessary to derive the performance metrics can be decreased by a factor of 100 or more without affecting the accuracy of the results produced.

Further, in 2016, R. D. Mohammady et al. [91] proposed a MAC layer CSMA/CA-based full-duplex analytical model for a single-cell wireless LAN network constituted of mobile servicing clients-based Access Points (APs). The authors first work out the Markov chain technique which yields closed-form throughput equations for clients and AP. After that, they use the full-duplex transmission to minimize the traditional hidden terminal problem. They also use star topology to form a network having a centre as AP and derived closed-form of successful transmission probability expressions and throughput expressions separately for the clients and AP. The proposed mechanism modifies the prior work on the MAC layer and analyses the uplink and downlink traffic flows from clients to the AP and AP to the client, respectively. The authors use *ns-2* based packet-level simulation to analyse the throughput of full-duplex and half-duplex networks and study the following points: (1) By using RTS/CTS in CSMA/CA, full-duplex transmission achieves up to 40% throughput improvement as compared to half-duplex transmission; (2) Changes to the Contention Window (CW) must be made with network size in mind, as there is a definite optimum gain for certain network sizes and contention window selection; (3) Full-duplex (FD) networks with a smaller CW size (about 128) are optimized around 25% better than half-duplex (HD) in an environment with significantly more concealed terminals that can be further optimized by increasing the size of CW; (4) The highest FD throughput and maximum FD gain can be achieved in the same CW so to design a network, improving the window size for anyone will be sufficient.

Further, in 2019, A. Maatouk et al. [92] proposed a new CSMA-based protocol that optimizes the throughput and energy efficiency of the CSMA protocol by modifying the power consumption and throughput requirements of each link. This scheme controls the sleeping and back-off timer's operational parameters to optimize the objective function. The authors introduce the distributed CSMA mechanism that avoids message passing and combines the energy efficiency and throughput aspects of the CSMA protocols. In this protocol, each link can switch between AWAKE and SLEEP states. The sleeping length of each link is not defined and is calibrated to optimize a specific target function. In this protocol, the authors adjust both the sleeping and back-off duration of each link in concert to optimize a certain objective function. Each link's power consumption may be decreased while still maintaining the ability to endure any possible arrival rate. As a consequence of the necessary study, the output will be a completely distributed MAC technique that is not time-slotted (thus requiring no synchronization) and is free of message passing and clock drift concerns. Due to the dynamic nature of link activity, this protocol achieves similar collision performance to adaptive CSMA while consuming significantly less power. The authors provide a computer-based simulation and compare the simulated results with IEEE 802.11 standard. The simulation results show that this scheme optimizes the throughput up to 20% and power consumption up to 15% as compared to IEEE 802.11 and adaptive CSMA, respectively.

11.6.8 SECURITY

In IoT, security creates challenges to protect the devices from attackers [6]. In IoT systems, it is not easy to guarantee security because billions of devices share the information having the same standard and architecture in the heterogeneous network [5]. Also, social IoT is allowing IoT devices to communicate with social networks [33]. Most IoT systems use wireless communication channels that can be easily hacked by attackers so there is a need for an end-to-end protection mechanism in IoT systems. Data transmission security and privacy are big issues for wireless sensors-based IoT networks due to attackers such as Man In the Middle Attacks (MIMAs), etc. [93].

In this regard, in 2017, M. Koseoglu et al. [94] analysed the energy consumption problem in underwater networks by using cross-layer (PHY-MAC)-based ALOHA protocols. The authors first obtained the maximum channel access rate to reduce the energy consumption per bit transmission for the ALOHA MAC layer. After that, the authors developed a cross-layer enhancement problem that works collaboratively on PHY and MAC layers to reduce power consumption. The authors look at a single-hop network in which nodes send data to a single base station. The channel is shared by N-saturated nodes meaning that nodes always have data to deliver. The nodes send the packet with a channel access rate using a Poisson process. The node resends the lost packet on the next transmission. The authors only analyse the traffic between nodes and the base station. For MAC layer optimization, the authors assume that nodes are equidistant and have the same transmission power. For cross-layer optimization (PHY and MAC layer), the authors look at how the probing rates and node's transmission powers in an ALOHA network are combined to reduce the network's energy usage. To analyse the proposed scheme, the authors use a large network (100 km radius) and a small network (10 km radius) for separate layer and cross-layer performance analysis. This mechanism allocates more resources (MAC layer) to the large distance nodes from the base station, resulting in an inefficient PHY layer. The numerical analysis presented in this chapter claims that this optimized cross-layer mechanism reduces the per-bit energy consumption up to 66% as compared to separate optimized physical and MAC layer.

Further, in 2019, F. Meneghello et al. [95] provided a survey for the most vulnerable attacks in IoT networks such as Dos attacks, unauthorized access and information leakage, and studied security mechanisms in different protocols used in IoT networks.

Further, in 2019, D. Shin et al. [96] proposed a secure route enhancement protocol based on distributed IP mobility management (DMM) to optimize security issues for smart home IoT applications. This protocol allows direct communication between devices and reduced the security attack in terms of data transmission information leakage. This protocol is a combination of handover phases and route optimization initialization to provide a key exchange, mutual authentication, privacy protection and perfect forward secrecy. The authors use BAN-logic and Automated Validation of Internet Security Protocols and Applications (AVISPA) tools to verify the proposed protocol and show that this protocol optimizes the security as compared to EAP-AKA, EAP-TLS and EAP-IKEv2 approaches.

Further, in 2020, A. Kore et al. [93] proposed an algorithm for wireless sensors-based IoT networks to detect middleman attackers called a cross-layer man-in-middle attack system (IC-MADS). This algorithm uses the clustering method to select Clusters Head (CH) and clusters formation by using sensor node parameters. The proposed algorithm is simulated on ns-2 and the authors show that the proposed algorithm (IC-MADS) is better in terms of attack detection as compared to MIMA.

Further, in 2020, A. Raghuprasad et al. [97] proposed an idea on how attackers hack information and how we can detect attacks such as air cracking for small-scale IoT networks. The proposed system has NodeMCU (microcontroller) hardware for the gateway, DHT11 sensors to measure the temperature, humidity and surrounding air, and MAC address to prevent the attack such as Denial of Service (DoS) and Distributed Denial of Service (DDoS). The proposed idea uses an additional security layer (MAC changer) to prevent the attack.

11.7 CONCLUSION

Depending on the architecture and technologies (such as Wi-Fi, WiMAX, Bluetooth, ZigBee, etc.) being used, the IoT protocol design will vary. In IoT, the protocol design mainly depends on four layers called physical, data link, application and network layer. To achieve a smart world with the full efficiency of IoT and to meet the application requirements, several research challenges such as energy consumption, scalability, reliability, mobility, interoperability, Quality-of-Service (QoS), system design, data usability and security challenges that affect the potential use of the IoT system must be optimized. In this comprehensive survey, a fundamental understanding of the IoT, its application, challenges and their state-of-the-art solution have been discussed. In this study, four layers of IoT architecture have been presented to describe the basic network requirement of the IoT system. Additionally, some emerging IoT applications and IoT features are discussed. Further, some IoT standards and the most appealing technologies such as ZigBee, RFID, Bluetooth, Wi-Fi, WiMAX, LoRaWAN, SIGFOX and NB-IoT have been provided. After that, some IoT challenges such as availability, reliability, scalability, mobility, privacy, security and interoperability, and their state-of-the-art solutions have provided seamless connectivity to fulfill IoT users' requirements.

REFERENCES

1. X. Cao, Z. Song, B. Yang, M. A. Elmossallamy, L. Qian, and Z. Han, "A distributed ambient backscatter MAC protocol for internet-of-things networks." *IEEE Internet of Things Journal*, 7(2), 2020, pp. 1488–1501, doi: 10.1109/JIOT.2019.2955909.
2. P. K. Verma, R. Verma, A. Parkash, A. Agrawal, K. Naik, R. Tripathi, M. Alsabaan, T. Khalifa, T. Abdelkader, and A. Abogharaf, "Machine-to-machine (M2M) communication: A survey." *Journal of Network and Computer Applications*, Elsevier, 2016, pp. 83–105, doi: 10.1016/j.jnca.2016.02.016.
3. P. K. Verma, R. Verma, A. Prakash, and R. Tripathi, "Massive access control in machine-to-machine communications." Algorithms, Methods, and Applications in Mobile Computing and Communications, IGI Global, 2019, pp. 133–157, doi: 10.4018/978-1-5225-5693-0.ch006.
4. J. Wang, M. K. Lim, C. Wang, and M. L. Tseng, "The evolution of the Internet of Things (IoT) over the past 20 years," *Computers and Industrial Engineering*, Elsevier, 155, 2021, p. 107174, doi: 10.1016/j.cie.2021.107174.
5. A. A. Fuqaha, M. Guizani, M. Mohammadi, M. Aledhari, and M. Ayyash, "Internet of things: A survey on enabling technologies, protocols, and applications." *IEEE Communications Surveys and Tutorials*, 17(4), 2015, pp. 2347–2376, doi: 10.1109/COMST.2015.2444095.
6. S. N. Swamy, and S. R. Kota, "An empirical study on system level aspects of Internet of Things (IoT)." *IEEE Access*, 8, 2020, pp. 188082–188134, doi: 10.1109/ACCESS.2020.3029847.
7. P. K. Verma, R. Verma, M. M. Alrayes, A. Prakash, R. Tripathi, and K. Naik, "A novel energy efficient and scalable hybrid-MAC protocol for massive M2M networks." *Cluster Computing*, Springer, 2018, pp. 1–22, doi: 10.1007/s10586-018-1948-y.
8. A. Rajandekar, and B. Sikdar, "A survey of MAC layer issues and protocols for machine-to-machine communications." *IEEE Internet of Things Journal*, 2(2), 2015, pp. 175–186, doi: 10.1109/JIOT.2015.2394438.
9. P. K. Verma, R. Verma, A. Prakash, and R. Tripathi, "Throughput enhancement of a novel hybrid-MAC protocol for M2M networks." *International Journal of Big Data Intelligence*, Inderscience Enterprises, 4(3), 2017, pp. 149–159.

10. A. E. Bouaouad, A. Cherradi, S. Assoul, and N. Souissi, "The key layers of IoT architecture." *5th International Conference on Cloud Computing and Artificial Intelligence: Technologies and Applications (CloudTech)*, 2020, pp. 1–4, doi: 10.1109/CloudTech49835.2020.9365919.

11. R. Khan, S. U. Khan, R. Zaheer, and S. Khan, "Future Internet: The Internet of things architecture, possible applications and key challenges." *10th International Conference on Frontiers of Information Technology*, 2012, pp. 257–260, doi: 10.1109/FIT.2012.53.

12. Z. Yang, Y. Yue, Y. Yang, Y. Peng, X. Wang, and W. Liu, "Study and application on the architecture and key technologies for IoT." *International Conference on Multimedia Technology*, 2011, pp. 747–751, doi: 10.1109/ICMT.2011.6002149.

13. P. Park, P. Di Marco, P. Soldati, C. Fischione, and K. H. Johansson, "A generalized Markov chain model for effective analysis of slotted IEEE 802.15.4." *IEEE 6th International Conference on Mobile Adhoc and Sensor Systems*, 2009, pp. 130–139, doi: 10.1109/MOBHOC.2009.5337007.

14. A. G. Khairnar, and D. A. Birari, "IoT ('connected life') and its use in different applications: A survey." *MVP Journal of Engineering Sciences*, 1(1), 2018, pp. 24–28, doi: 10.18311/mvpjes/2018/v1i1/19962.

15. N. Barodawala, B. Makwana, Y. Punjabi, and C. Bhatt, *Home automation using IoT.* Springer, New York, NY, 2018, pp. 219–242.

16. A. Tiele, S. Esfahani, and J. Covington, "Design and development of a low-cost, portable monitoring device for indoor environment quality." *Journal of Sensors*, 2018, pp. 1–14, doi: 10.1155/2018/5353816.

17. B. Min, and S. J. Park, "A smart indoor gardening system using IoT technology." *Lecture Notes in Computer Science*, 474, 2018, pp. 683–687, doi: 10.1007/978-981-10-7605-3-110.

18. A. Based, I. Froiz-Míguez, T. M. Fernández-Caramés, P. Fraga-Lamas, and L. Castedo, "Design, Implementation and Practical Evaluation of an IoT Home Automation System for Fog Computing Applications Based on MQTT and ZigBee-WiFi Sensor Nodes." *Sensors*, 18(8), 2018, pp. 1–42.

19. A. Abdelgawad, K. Yelamarthi, and A. Khattab, "IoT-based health monitoring system for active and assisted living." *Proceedings of the Lecture Notes of the Institute for Computer Sciences*, 2017, pp. 11–20, doi: 10.1007/978-3-319-61949-1-2.

20. S. Velten, J. Leventon, N. Jager, and J. Newig, "What is sustainable agriculture? A systematic review." *Sustainability*, 7(6), 2015, pp. 7833–7865, doi: 10.3390/su7067833.

21. S. Banu, "Precision agriculture: Tomorrow's technology for today's farmer." *Food Process. Technol*, 6(8), 2015, pp. 8–13, doi: 10.4172/2157-7110.1000468.

22. A. Zear, P. K. Singh, and Y. Singh, "Intelligent transport system: A progressive review." *Indian Journal of Science and Technology*, 9(32), 2016, doi: 10.17485/ijst/2016/v9i32/100713.

23. D. Jaisinghani, P. A. M. Bongale, and G. H. Raisoni, "Real-time intelligent traffic light monitoring and control system to predict traffic congestion using data mining and WSN." *International Journal of Engineering Research And*, V4(12), 2015, pp. 33–38.

24. A. Badri, B. Boudreau-Trudel, and A. S. Souissi, "Occupational health and safety in the industry 4.0 era: A cause for major concern?" *Safety Science*, Elsevier, 109, 2018, pp. 403–411, doi: 10.1016/j.ssci.2018.06.012.

25. H. Lasi, P. Fettke, H. G. Kemper, T. Feld, and M. Hoffmann, "Industry 4.0." *Business and Information Systems Engineering*, 6(4), 2014, pp. 239–242, doi: 10.1007/s12599-014-0334-4.

26. K. Santos, E. Loures, F. Piechnicki, and O. Canciglieri, "Opportunities assessment of product development process in industry 4.0." *Procedia Manufacturing*, 11, 2017, pp. 1358–1365, doi: 10.1016/j.promfg.2017.07.265.

27. M. A. A. Mamun, and M. R. Yuce, "Sensors and systems for wearable environmental monitoring toward IoT-enabled applications: A review." *IEEE Sensors Journal*, 19(18), 2019, pp. 7771–7788, doi: 10.1109/JSEN.2019.2919352.

28. G. Mois, S. Folea, and T. Sanislav, "Analysis of three IoT-based wireless sensors for environmental monitoring." *IEEE Transactions on Instrumentation and Measurement*, 66(8), 2017, pp. 2056–2064, doi: 10.1109/TIM.2017.2677619.

29. P. Kaur, M. Singh, and G. S. Josan, "Classification and prediction based data mining algorithms to predict slow learners in education sector." *3rd International Conference on Recent Trends in Computing (ICRTC-2015)*, 2015, pp. 500–508.

30. S.-C. Hsia, Y.-J. Chang, and S.-W. Hsu, "Remote monitoring and smart sensing for water meter system and leakage detection." *IET Wireless Sensor Systems*, 2(4), 2012, pp. 402–408, doi: 10.1049/iet-wss.2012.0062.

31. A. Sinha, P. Kumar, N. P. Rana, R. Islam, and Y. K. Dwivedi, "Impact of Internet of Things (IoT) in disaster management: A task-technology fit perspective." *Annals of Operations Research*, 283(1–2), 2019, pp. 759–794, doi: 10.1007/s10479-017-2658-1.

32. A. J. Meijer, J. R. Gil-Garcia, and M. P. R. Bolívar, "Smart city research." *Social Science Computer Review*, 34(6), 2016, pp. 647–656, doi: 10.1177/0894439315618890.

33. R. I. Ogie, P. Perez, and V. Dignum, "Smart infrastructure: An emerging frontier for multidisciplinary research." *Proceedings of the Institution of Civil Engineers - Smart Infrastructure and Construction*, 170(1), 2017, pp. 8–16, doi: 10.1680/jsmic.16.00002.

34. S. Durga, S. Surya, and E. Daniel, "SmartMobiCam: Towards a new paradigm for leveraging smartphone cameras and IaaS cloud for smart city video surveillance." *Proceedings 2nd International Conference Trends Electron, Information (ICOEI)*, 2018, pp. 1035–1038, doi: 10.1109/ICOEI.2018.8553974.

35. B. N. Silva, M. Khan, and K. Han, "Towards sustainable smart cities: A review of trends, architectures, components, and open challenges in smart cities." *Sustainable Cities and Society*, 38, 2018, pp. 697–713, doi: 10.1016/j.scs.2018.01.053.

36. M. A. Adegboye, W.-K. Fung, and A. Karnik, "Recent advances in pipeline monitoring and oil leakage detection technologies: Principles and approaches." *Sensors*, 19(11), 2019, p. 2548, doi: 10.3390/s19112548.

37. Z. Chen, X. Zhou, X. Wang, L. Dong, and Y. Qian, "Deployment of a smart structural health monitoring system for long-span arch bridges: A review and a case study." *Sensors*, 17(9), 2017, p. 2151, doi: 10.3390/s17092151.

38. J. C. Talwana, and H. J. Hua, "Smart world of internet of things (IoT) and its security concerns," *2016 IEEE International Conference on Internet of Things (iThings) and IEEE Green Computing and Communications (GreenCom) and IEEE Cyber, Physical and Social Computing (CPSCom) and IEEE Smart Data (SmartData)*, Chengdu, China, 2016, pp. 240–245, doi: 10.1109/iThings-GreenCom-CPSCom-SmartData.2016.64.

39. E. Ferro, and F. Potorti, "Bluetooth and wi-fi wireless protocols: A survey and a comparison." *IEEE Wireless Communications*, 12(1), 2005, pp. 12–26, doi: 10.1109/MWC.2005.1404569.

40. "Press releases detail: Bluetooth technology website." *Bluetooth Technology*. Website, Kirkland, WA, 2014, https://www.bluetooth.com/.

41. R. Want, "An introduction to RFID technology." *IEEE Pervasive Computing*, 5(1), 2006, pp. 25–33, doi: 10.1109/MPRV.2006.2.

42. P. Barnaghi, W. Wang, C. Henson, and K. Taylor, "Semantics for the Internet of Things: Early progress and back to the future." *Proceedings of the International Journal on Semantic Web and Information Systems*, 8(1), 2012, pp. 1–21.

43. T. Kamiya, and J. Schneider, "Efficient XML interchange (EXI) Format 1.0." *World Wide Web Consortium, Cambridge, MA, Recommend*. REC-Exi-20110310, 2011.

44. T. Perumal, S. K. Datta, and C. Bonnet, "IoT device management framework for smart home scenarios." *2015 IEEE 4th Global Conference on Consumer Electronics, GCCE 2015*, 2015, pp. 54–55, doi: 10.1109/GCCE.2015.7398711.

45. IETF Working Groups, "2C spanning build and specify interoperable systems IETF. The Internet of things." 2020 [Online]. Available: https://www.ietf.org/topics/iot/.

46. A. Pal, H. K. Rath, S. Shailendra, and A. bhattacharya, "IoT standardization: The road ahead." *Internet of Things - Technology, Applications and Standardization*, 2018, pp. 53–74, doi: 10.5772/intechopen.75137.

47. C. M. Ramya, M. Shanmugaraj, and R. Prabakaran, "Study on Zig-Bee technology." *Proceedings of the 3rd International Conference on Electronic Computing Technology*, 2011, pp. 297–301, doi: 10.1109/ICECTECH.2011.5942102.

48. "Wireless medium access control and physical layer specifications for low-rate wireless personal area networks." 2003. *IEEE Standard 802.15.4*.

49. ZigBee Specification, Revision 20, document 053474r20, ZigBee Alliance, Davis, CA, 2012.

50. L. Chettri, and R. Bera, "A comprehensive survey on Internet of Things (IoT) toward 5G wireless systems." *IEEE Internet of Things Journal*, 7(1), 2020, pp. 16–32, doi: 10.1109/JIOT.2019.2948888.

51. K. E. Jeon, J. She, P. Soonsawad, and P. C. Ng, "BLE beacons for Internet of Things applications: Survey, challenges, and opportunities." *IEEE Internet of Things Journal*, 5(2), 2018, pp. 811–828, doi: 10.1109/JIOT.2017.2788449.

52. L. Li, H. Xiaoguang, C. Ke, and H. Ketai, "The applications of WiFibased wireless sensor network in Internet of Things and smart grid." *2011 6th IEEE Conference on Industrial Electronics and Applications*, 2011, pp. 789–793.

53. J. Kim, and J. Song, "A secure device-to-device link establishment scheme for LoRaWAN." *IEEE Sensors Journal*, 18(5), 2018, pp. 2153–2160, doi: 10.1109/JSEN.2017.2789121.

54. H. Mroue, A. Nasser, S. Hamrioui, B. Parrein, E. Motta-Cruz, and G. Rouyer, "MAC layer-based evaluation of IoT technologies: LoRa, SigFox and NB-IoT." *IEEE Middle East and North Africa Communications Conference (MENACOMM)*, 2018, pp. 1–5, doi: 10.1109/MENACOMM.2018.8371016.

55. K. Mekki, E. Bajic, F. Chaxel, and F. Meyer, "Overview of cellular LPWAN technologies for IoT deployment: Sigfox, LoRaWAN, and NBIoT." *Proceedings of the IEEE International Conference on Pervasive Computing and Communications Workshops (PerCom Workshops)*, 2018, pp. 197–202.

56. A. Lavric, A. I. Petrariu, and V. Popa, "Long range SigFox communication protocol scalability analysis under large-scale, high-density conditions." *IEEE Access*, 7, 2019, pp. 35816–35825, doi: 10.1109/ACCESS.2019.2903157.

57. "Sigfox technology overview." [Online]. Available: https://www.sigfox.com/en/sigfox -iot-technology-overview.

58. C. B. Mwakwata, H. Malik, M. M. Alam, Y. Le Moullec, S. Parand, and S. Mumtaz, "Narrowband Internet of Things (NB-IoT): From physical (PHY) and media access control (MAC) layers perspectives." *Sensors*, 1(11), 2019, pp. 1–34, doi: 10.3390/s19112613.

59. Y. Li, X. Cheng, Y. Cao, D. Wang, and L. Yang, "Smart choice for the smart grid: Narrowband Internet of Things (NB-IoT)." *IEEE Internet of Things Journal*, 5(3), 2018, pp. 1505–1515, doi: 10.1109/JIOT.2017.2781251.

60. J. Silva, J. Rodrigues, J. Al-Muhtadi, R. Rabelo, and V. Furtado, "Management platforms and protocols for Internet of Things: A survey." *Sensors*, 19(3), 2019, p. 676, doi: 10.3390/s19030676.

61. C. Perera, A. Zaslavsky, P. Christen, and D. Georgakopoulos, "Context aware computingfor the Internet of Things: A survey." *IEEE Communications Surveys and Tutorials*, 16(1), 2013, pp. 414–454, doi: 10.1109/SURV.2013.042313.00197.

62. D. Macedo, L. A. Guedes, and I. Silva, "A dependability evaluation for Internet of Things incorporating redundancy aspects." *Proceedings of the 11th IEEE International Conference on Networking, Sensing and Control*, 2014, pp. 417–422, doi: 10.1109/ICNSC.2014.6819662.

63. Q.Wei, Z. Jin, L. Li, and G. Li, "Lightweight semantic service modeling for IoT: An environment-based approach." *International Journal of Embedded Systems*, 8(2–3), 2016, pp. 164–173, doi: 10.1504/IJES.2016.076110.

64. W. B. Qaim, and O. Ozkasap, "DRAW: Data replication for enhanced data availability in IoT-based sensor systems." *IEEE 16th International Conference on Dependable, Autonomic and Secure Computing, 16th International Conference on Pervasive Intelligence and Computing, 4th International Conference on Big Data Intelligence and Computing and Cyber Science and Technology Congress(DASC/PiCom/DataCom/CyberSciTech)*, 2018, pp. 770–775, doi: 10.1109/DASC/PiCom/DataCom/CyberSciTec.2018.00133.

65. M. A. Razzaque, M. M. Jevric, A. Palade, and S. Clarke, "Middleware for Internet of things: A survey." *IEEE Internet of Things Journal*, 3(1), 2016, pp. 70–94, doi: 10.1109/JIOT.2015.2498900.

66. J. Kempf, J. Arkko, N. Beheshti, and K. Yedavalli, "Thoughts on reliability in the Internet of Things." *Proceedings of the Interconnecting Smart Objects Internet Workshop*, 2011, pp. 1–4.

67. M. Zayani, V. Gauthier, and D. Zeghlache, "A joint model for IEEE 802.15.4 physical and medium access control layers." *7th International Wireless Communications and Mobile Computing Conference*, 2011, pp. 814–819, doi: 10.1109/IWCMC.2011.5982651.

68. M. P. R. S. Kiran, V. Subrahmanyam, and P. Rajalakshmi, "Novel power management scheme and effects of constrained on-node storage on performance of MAC layer for industrial IoT networks." *IEEE Transactions on Industrial Informatics*, 14(5), 2018, pp. 2146–2158, doi: 10.1109/TII.2017.2766783.

69. Y. Jin, U. Raza, A. Aijaz, M. Sooriyabandara, and S. Gormus, "Content centric cross-layer scheduling for industrial IoT applications using 6TiSCH." *IEEE Access*, 6, 2018, pp. 234–244, doi: 10.1109/ACCESS.2017.2762079.

70. A. Bakshi, L. Chen, K. Srinivasan, C. E. Koksal, and A. Eryilmaz, "EMIT: An efficient MAC paradigm for the Internet of things." *IEEE/ACM Transactions on Networking*, 27(4), 2019, pp. 1572–1583, doi: 10.1109/TNET.2019.2928002.

71. F. Li, M. Voegler, M. Claessens, and S. Dustdar, "Efficient and scalable IoT service delivery on cloud." *2013 IEEE Sixth International Conference on Cloud Computing*, 2013, pp. 740–747, doi: 10.1109/CLOUD.2013.64.

72. C. Sarkar, S. N. A. U. Nambi, R. V. Prasad, and A. Rahim, "A scalable distributed architecture towards unifying IoT applications." *IEEE World Forum on Internet of Things (WF-IoT)*, 2014, pp. 508–513, doi: 10.1109/WF-IoT.2014.6803220.

73. N. Shahin, R. Ali, and Y. Kim, "Hybrid slotted-CSMA/CA-TDMA for efficient massive registration of IoT devices." *IEEE Access*, 6, 2018, pp. 18366–18382, doi: 10.1109/ACCESS.2018.2815990.

74. P. K. Verma, R. Verma, A. Prakash, and R. Tripathi, "Throughput-delay evaluation of a hybrid-MAC protocol for M2M communications." *International Journal of Mobile Computing and Multimedia Communications*, 7(1), 2016, pp. 41–60, doi: 10.4018/ijmcmc.2016010104.

75. P. K. Verma, R. Verma, A. Prakash, R. Tripathi, and K. Naik, "A novel hybrid medium access control protocol for inter M2M communications." *Journal of Network and Computer Applications*, Elsevier, 2016, pp. 77–88, doi: 10.1016/j.jnca.2016.08.011.

76. A. Bröring, S. Schmid, C.-K. Schindhelm, A. Khelil, S. Käbisch, D. Kramer, D. Le Phuoc, J. Mitic, D. Anicic, and E. Teniente, "Enabling IoT ecosystems through platform interoperability." *IEEE Software*, 34(1), 2017, pp. 54–61, doi: 10.1109/MS.2017.2.

77. S. Yang, and R. Wei, "Tabdoc approach: An information fusion method to implement semantic interoperability between IoT devices and users." *IEEE Internet of Things Journal*, 6(2), 2019, pp. 1972–1986, doi: 10.1109/JIOT.2018.2871274.

78. P. K. Verma, A. Prakash, R. Tripathi, R. Verma, and K. Naik, "A novel scalable hybrid-MAC protocol for densely deployed M2M networks." *2015 International Conference on Computational Intelligence and Communication Networks*, IEEE, 2015, pp. 50–56, doi: 10.1109/CICN.2015.19.

79. P. K. Verma, R. Tripathi, and K. Naik, "A robust hybrid-mac protocol for M2M communications." *International Conference on Computer and Communication Technology (ICCCT)*, 2014, pp. 267–271.

80. H. Park, "Adaptive Backoff enabled WUR on non-cellular local IoT for extreme low power operation." *Future Generation Computer Systems*, Elsevier, 108, 2020, pp. 62–67.

81. M. P. R. S. Kiran, and P. Rajalakshmi, "Performance analysis of CSMA/CA and PCA for time critical industrial IoT applications." *IEEE Transactions on Industrial Informatics*, 14(5), 2018, pp. 2281–2293, doi: 10.1109/TII.2018.2802497.

82. X. Lin, G. Su, B. Chen, H. Wang, and M. Dai, "Striking a balance between system throughput and energy efficiency for UAV-IoT systems." *IEEE Internet of Things Journal*, 6(6), 2019, pp. 10519–10533, doi: 10.1109/JIOT.2019.2939823.

83. N. Choudhury, R. Matam, M. Mukherjee, and J. Lloret, "A performance-to-cost analysis of IEEE 802.15.4 MAC with 802.15.4e MAC modes." *IEEE Access*, 8, 2020, pp. 41936–41950, doi: 10.1109/ACCESS.2020.2976654.

84. O. Said, Z. Al-Makhadmeh, and A. Tolba, "EMS: An energy management scheme for green IoT environments." *IEEE Access*, 8, 2020, pp. 44983–44998, doi: 10.1109/ACCESS.2020.2976641.

85. B. Safaei, A. M. H. Monazzah, and A. Ejlali, "ELITE: An elaborated cross-layer RPL objective function to achieve energy efficiency in internet-of-things devices." *IEEE Internet of Things Journal*, 8(2), 2021, pp. 1169–1182, doi: 10.1109/JIOT.2020.3011968.

86. D. Wu, X. Nie, E. Asmare, D. I. Arkhipov, Z. Qin, R. Li, J. A. McCann, K. Li, "Towards distributed SDN: Mobility management and flow scheduling in software defined urban IoT." *IEEE Transactions on Parallel and Distributed Systems*, 31(6), 2020, pp. 1400–1418, doi: 10.1109/TPDS.2018.2883438.

87. K. Cao, G. Xu, J. Zhou, T. Wei, M. Chen, and S. Hu, "QoS-adaptive approximate real-time computation for mobility-aware IoT lifetime optimization." *IEEE Transactions on Computer-Aided Design of Integrated Circuits and Systems*, 38(10), 2019, pp. 1799–1810, doi: 10.1109/TCAD.2018.2873239.

88. A. A. R. Alsaeedy, and E. K. P. Chong, "Mobility management for 5G IoT devices: Improving power consumption with lightweight signaling overhead." *IEEE Internet of Things Journal*, 6(5), 2019, pp. 8237–8247, doi: 10.1109/JIOT.2019.2920628.

89. D. Striccoli, G. Boggia, and L. A. Grieco, "A markov model for characterizing IEEE 802.15.4 MAC layer in noisy environments." *IEEE Transactions on Industrial Electronics*, 62(8), 2015, pp. 5133–5142, doi: 10.1109/TIE.2015.2403792.

90. D. D. Guglielmo, F. Restuccia, G. Anastasi, M. Conti, and S. K. Das, "Accurate and efficient modeling of 802.15.4 unslotted CSMA/CA through event chains computation." *IEEE Transactions on Mobile Computing*, 15(12), 2016, pp. 2954–2968, doi: 10.1109/TMC.2016.2528248.

91. A. Maatouk, M. Assaad, and A. Ephremides, "Energy efficient and throughput optimal CSMA scheme." *IEEE/ACM Transactions on Networking*, 27(1), 2019, pp. 316–329, doi: 10.1109/TNET.2019.2891018.

92. R. D. Mohammady, M. Y. Naderi, and K. R. Chowdhury, "Performance analysis of CSMA/CA based medium access in full duplex wireless communications." *IEEE Transactions on Mobile Computing*, 15(6), 2016, pp. 1457–1470, doi: 10.1109/TMC.2015.2462832.

93. A. Kore, and S. Patil, "Robust cross-layer security framework for Internet of things enabled wireless sensor networks." *International Conference on Emerging Smart Computing and Informatics (ESCI)*, 2020, pp. 142–147, doi: 10.1109/ESCI48226.2020.9167555.

94. M. Koseoglu, E. Karasan, and L. Chen, "Cross-layer energy minimization for underwater ALOHA networks." *IEEE Systems Journal*, 11(2), 2017, pp. 551–561, doi: 10.1109/JSYST.2015.2475633.

95. F. Meneghello, M. Calore, D. Zucchetto, M. Polese, and A. Zanella, "IoT: Internet of Threats? A survey of practical security vulnerabilities in real IoT devices." *IEEE Internet of Things Journal*, 6(5), 2019, pp. 8182–8201, doi: 10.1109/JIOT.2019.2935189.

96. D. Shin, K. Yun, J. Kim, P. V. Astillo, J. Kim, and I. You, "A security protocol for route optimization in DMM-based smart home IoT networks." *IEEE Access*, 7, 2019, pp. 142531–142550, doi: 10.1109/ACCESS.2019.2943929.

97. A. Raghuprasad, S. Padmanabhan, M. Arjun Babu, and P. K. Binu, "Security analysis and prevention of attacks on IoT devices." *International Conference on Communication and Signal Processing (ICCSP)*, 2020, pp. 0876–0880, doi: 10.1109/ICCSP48568.2020.9182055.

98. S. Ravidas, A. Lekidis, F. Paci, and N. Zannone, "Access control in Internet-of-Things: A survey." *Journal of Network and Computer Applications*, 144, 2019, pp. 79–101.

99. Y. Lu, and L. D. Xu, "Internet of Things (IoT) cybersecurity research: A review of current research topics." *IEEE Internet of Things Journal*, 6(2), 2019, pp. 2103–2115, doi: 10.1109/JIOT.2018.2869847.

100. A. Nauman, Y. A. Qadri, M. Amjad, Y. B. Zikria, M. K. Afzal, and S. W. Kim, "Multimedia Internet of Things: A comprehensive survey." *IEEE Access*, 8, 2020, pp. 8202–8250, doi: 10.1109/ACCESS.2020.2964280.

101. V. Sharma, I. You, K. Andersson, F. Palmieri, M. H. Rehmani, and J. Lim, "Security, privacy and trust for smart Mobile-Internet of Things (M-IoT): A survey." *IEEE Access*, 8, 2020, pp. 167123–167163, doi: 10.1109/ACCESS.2020.3022661.

12 Cloud-based Ecosystem of IoMT

*Sai Hemanth Talluri, Pavan Sai Gopal
Kande, Ram Koushik Gopina, Rayala
Sai Abhi, and Yogesh Tripathi*

12.1 INTRODUCTION AND ORIGINS

The Internet of Things (IoT) has started to gain more popularity these days. In 2022, the market for the IoT is anticipated to grow by 18%, i.e., to 14.4 billion vigorous connections. By the end of 2025, there will be in the neighbourhood 27 billion connected IoT devices [1]. The Internet of Medical Things (IoMT) is a network of interconnected devices that are intended to transfer patients' health matrices like Electromyography (EMG), Electrocardiogram (ECG), feverishness, heart speed, Blood Pressure (BP), Arterial Oxygen Saturation (SpO2), etc. with the help of Internet to the healthcare service provider. The recent pandemic (Covid-19) has proven the fact that the healthcare industry is facing a shortage of healthcare staff, IT Healthcare Investments, etc. By 2025, there is a contemplated gap of 200,000 to 450,000 enlisted nursemaids and 50,000 to 80,000 medics (10% to 20% and 6% to 10% of the manpower, respectively) [2]. These contemporary problems can be addressed by the effective implementation of IoMT in the healthcare system. The cloud has proven its capacity on large scale. Cloud computing furnishes calculating benefits like databases or storage services, servers facility, software services, data analytics and networking over the Internet to furnish multifaceted aids, accelerated deployment and providence of scale [3]. The usage of the cloud in the transmission of such important data by maintaining confidentiality and integrity with proper authentication becomes more important. The cloud can be used for processing the data compiled by IoT devices and securely storing the data.

Surveys and reviews on the Ecosystem of IoMT have been done before by various researchers across the globe. The contributions of this research chapter are as follows:

1. It explains the basic working of the cloud-based IoMT.
2. Identifies a set of challenges that are associated with the cloud-based IoMT ecosystem.
3. Identifies the possible attacks and their effect on the data and lists some of the basic countermeasures for the attacks.

DOI: 10.1201/9781003452645-12

This chapter is arranged as follows: The Introduction provides basic insights of IoMT followed by a literature review which talks about previous works on this area. This is followed by classification of the IoMT which is followed by cloud-based IoMT. Finally, challenges in the IoMT is followed by attacks on the IoMT and future scope and conclusion.

12.2 LITERATURE REVIEW

The IoMT has been a notable area for experimenters who are functioning in the domain of the IoT, Artificial Intelligence (AI), Machine Learning (ML), Cloud, and Cyber Security. A Survey on Security Dangers and Countermeasures in the IoMT [4], has given comprehensive insights into the security aspects of the IoMT from the basic CIA triad (Confidentiality, Integrity and Availability) to the non-exclusive attack kind in the IoMT edge network. In IoMT, the data are sensitive and Personal Identifiable Information (PII) should be handled very carefully. This data should be authentic, confidential and be available with integrity. The security requirements of the industry and government becoming more complex day by day. The challenges associated with identifying and detecting insider threats becomes more complex. The mitigation of the threats by the malicious insider will directly be dependent on behavioural and technical solutions [47]. When we are dealing with the insider threat, access to critical data plays a significant role in maintaining integrity, confidentiality and availability of critical data. So, the data generated manually don't provide consistency and coverage to test complex systems and this comes with a high rate of error and price. The generation of data sets is static and limited in supply. This makes it very hard to get considerable information sets that make a match with known characteristics; thus data sets cannot be disseminated without jeopardizing classified information and being personally identifiable [46]. For distributed network attacks like Distributed Denial of Service (DDoS) attacks, the Collaborative Network Security Management System (CNSMS) can be employed as a countermeasure for broadcasted network incursions. Unified Threat Management (UTM) is developed for large volume of data from the CNSMS. This includes security events, network traffics, etc. The collected data can be used in a cloud computing system to track the attacking events [50]. There are many mechanisms that have been developed against DDoS attacks, but the effectiveness of these mechanisms has always remained a concern because these are not always effective. The different mechanisms to construct holistic hybrid defence mechanisms will be a solution to improve the effectiveness like Ip Spoofing-recommended defence mechanism and Hop Count Filtering (HCF) in the PaaS layer. In an SYN flooding attack, the SYN cookies protection procedure can be employed in the PaaS layer or diminish the duration of SYN received in the PaaS layer. In the circumstance of a Smurf attack, the configuring of virtual machines in the PaaS layer or network aids in the IaaS layer mechanisms. In the case of Buffer overflow avoidance, when writing source code mechanism or runtime instrumentation mechanism can be used in the ping of death attack, Land.c, and Teardrop.c [63]. When it is a security concern in the Smart Home or WSN or UAWSN, there are a few existing solutions to the existing attacks for jamming attacks. By encrypting the message, we can prevent tampering attacks on the Smart Home using Code Division Multiple Access (CDMA). Also, there is an attack called a blackhole attack that can be evaded by updating the routing tables in WSN or Mobile UWSN. In VANET, this attack can be evaded by checking the packet sequence numbers from the packet header [31]. A blackhole attack can significantly drop the performance of the network. So, schemes like root-based defence schemes detect malicious nodes in the blackhole by sending a packet loss detection algorithm to the root node. This will publicize blackhole node details to the entire network. This broadcasted information is used by the non-root nodes to avoid or isolate the blackhole nodes [32]. The conventional security protocols are not satisfactory for the IoMT because of its multifariousness in the hardware proficiency of objects implicated in multimedia communication. With the proposal of a cross-layer communication of the layered stack system has solved problems that were traditional in the communication protocol [5]. The idea of integrating blockchain into authentication key agreement protocol for the IoMT has shown how to provide secure key management among different communicating entities [6]. The data can be altered in transit while pushing from edge to cloud by employing the Man In The Middle (MITM) attacks in the PhotoVoltaic PV systems. This danger can be handled by blockchain-based MITM attack detection. This approach uses the shield module that is concatenated with Operational Technology (OT) devices in PhotoVoltaic (PV) systems. This furnishes the logs of the crucial aids which operate as zero-trust systems for PV systems [23]. But it is needed to tackle the security issues of the back door data in the edge services. So, security management in edge intelligent services can be improved by detecting and eliminating backdoor data. The self-attention distillation will be steered on the revamped model, and which is transmitted to the cloud server to enhance the accuracy and protection of the model. After achieving the self-attention distillation, the exactness is enhanced by 2% and this is comparable to the performance of the clean model [29]. In account hijacking, hackers typically endeavour to utilize the compromised email account. Majorly account hijacking is brought out via phishing or mailing spoofed emails to the target or attempting to assume the passwords. In many cases, users or victims use the same email id or passwords on various other sites if any other site is compromised by any other attack, then it leads to access the financial transactions, personal information, etc. In the cloud, it is a common tactic to perform account hijacking by theft techniques. The assaulter typically uses the pinched knowledge to accomplish unauthorized and adversarial activity [52]. In the case of the *New York Times* (NYT), their website was taken down for almost 6 hours. This was due to the poor implementation of security policy in Melbourne IT. To perform the attack on the NYT, the attacker underwent a detailed study of the NYT hosting and IT infrastructure. So, cloud service providers should take action against the policy of sharing account credentials between clients and service providers [34]. The cloud is being adopted rapidly in the business world, but the possibility of a zombie attack has remained a concern in the cloud. This attack leads to a decrease in the performance of the network in terms of both usage and delay in the speed of the network [40]. To deal with the attacks on the cloud, i.e., denial-of-service attacks we should filter and monitor the traffic. Cloud providers like Amazon AWS provide Amazon Shield and Cloudflare which can be used to defend against DoS and DDoS attacks by forwarding only requests originating from legitimate sources and blocking other requests. Multi-Factor Authentication (MFA) should be used to protect from identity theft. Data Loss Prevention (DLP) can be employed to contain data loss in the cloud and bypass unauthorized handling or leakage of data in the cloud [6]. Here in this procedure, the entire healthcare data are stored in a blockchain maintained by the cloud servers. With the introduction of 5G, the IoMT in health care saw better optimistic methods to trade with remote treatment for patients. As a result, problems in the IoMT ecosystem have decreased [57].

12.3 CLASSIFICATION OF THE INTERNET OF MEDICAL THINGS

12.3.1 IN-HOME IoMT

With In-home IoMT, the health data are transferred from home to licensed primary healthcare centres, specialized hospitals and analysis institutes. Remote Patient Observance (RPM) is the primary example of In-home IoMT. The market capitalization rate is anticipated to succeed in US$ 2114 million by 2028 [7]. From recently discharged patients In-home IoMT collects health matrices and transmits them to the hospital which will be reviewed by their doctors to provide the required treatment. This will scale down hospital readmissions by catching problems before they turn out to be serious.

12.3.2 ON-BODY IoMT

On-body IoMT are classified as devices worn on the body of the patient, known as the Body Area Network (BAN), so the patient can carry them to workplaces, schools, colleges, etc. Also, data can be sent in real-time to the healthcare centers or we can use this data to track personal health matrices.

12.3.3 COMMUNITY IoMT

When the health data are collected for a group of people across society, towns, etc. it is known as community IoMT. This involves remote services like when a patient is in transit from home to hospital or vice versa then devices are used to track patient health data. When medical equipment is in transit then we can use the IoMT to track the condition of the medical equipment using sensors that track temperature, pressure, etc., for example, in a shipping container that ideally needs to maintain for the proper working of medical devices.

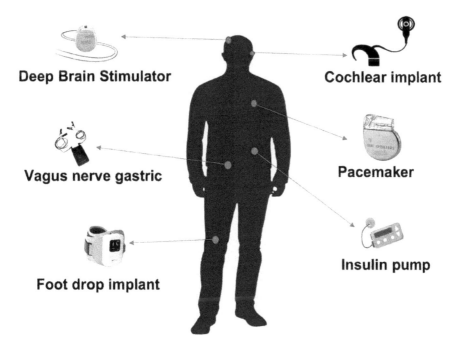

Deep Brain Stimulator

Cochlear implant

Vagus nerve gastric

Pacemaker

Foot drop implant

Insulin pump

FIGURE 12.1   On body network of the IoMT [105]

12.3.4   IN-HOSPITAL IoMT

Hospitals across the globe must manage their quality of services, and time spent on the individual patient for better treatment of the patient. In the hospital, we can use the IoMT devices as shown in Figure 12.1 to cautiously monitor the health condition of patients which helps to improve the quality of treatment for the patient.

12.4   CLOUD-BASED IOMT

When a patient gets discharged from the hospital, doctors need to monitor the patient's recovery which is possible via cloud-based IoMT. The early outpatient fol-low-up does not benefit the patients where 9% to 55% of total readmission in hospi-tals is due to improper healthcare monitoring after discharge from the hospital and by inappropriate care during the index admission [8, 9]. In Figure 12.2, the IoMT eco-system can be used to monitor patient health data.

In IoMT Echo-system, we can use on-body sensors which are connected to the Internet and will dispatch the information securely to the cloud. These devices include smart watches, phones, etc. This request made by on-body sensors and other connected devices are filtered by the cloud firewall. This filter are which is generated

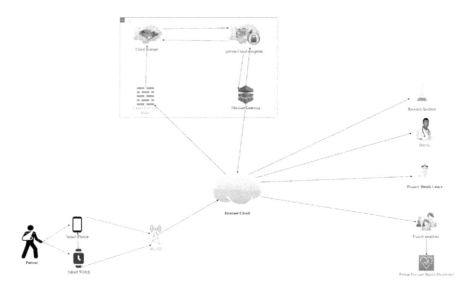

FIGURE 12.2 Cloud-based ecosystem of IoMT

by the user behaviour rules. Major cloud service provides cloud firewall rules like Google VPC firewall rules, AWS Network Firewall, and Azure Firewall [10–12]. Then the information is stowed in the cloud storage services. Based on the data sent to the cloud, the computing services run the Machine Learning algorithm. This Machine Learning algorithm classifies the data and sends an alert to the primary healthcare centre, doctors, patients' health dashboard and family members. Based on this continuous monitoring the doctors make a change to the treatment that will help improve the quality of healthcare.

12.5   CHALLENGES IN THE IOMT

When data are transmitted through the Internet there is always risk associated with it. We need to tackle this risk by evaluating the nature of the risk or challenges in transmitting, storing and processing the data. The nature of data handled in IoMT is too sensitive. Also we need to evaluate information risk above 12.5.1[103].

12.5.1   DATA LOSS AND LEAKAGE

It is a direct attack on the availability, confidentiality and authenticity of critical health data. These data are so sensitive if the data are lost or modified that it will impact the purpose of the IoMT ecosystem. So, there should be an original backup of the data. The loss of the encryption key also results in data loss. For example, consider a user suffering from irregular BP. If the data are altered by a malicious entity which results in wrong treatment by doctors then it leads to severe consequences. Data loss and data leaks are caused by weak authentication, access controls and authorizations including the inability of data centers to recover from natural disasters. Using secure API, secure storage, data backups and encryption algorithms and keys reduce the risk of data loss and leakage.

12.5.2 INSECURE API AND INTERFACES

In the cloud, the communication between cloud services is done using API and interfaces. These interfaces are in the form of a layer, so this increases the complication of the cloud. This interface contributes to management and monitoring assistance. So, the protection of the cloud is instantly affected by the safety of the API. If the API is insecure then attackers may use this as an opportunity to gain unauthorized access to the data and affect the integrity of data by altering it. This can be prevented by acquiring secure authentication and access control mechanisms and secure interface models.

12.5.3 MALICIOUS INSIDER

It is a major threat because many healthcare centers overlook the access level of internal aid for their workers. If somehow any employee gets a more elevated level of permit that results in an infringement of confidentiality of the data this can be watched by the intruder detection system based on analysing the behavioural routines of the individual and term it as an intruder or not. This helps us to avoid major risks. This malicious person can modify the existing data in the system and eventually cause huge risks.

12.5.4 ACCOUNT HIJACKING

It is a method of collecting information or credentials from the victim via fraud, phishing and exploitable software vulnerabilities. The reuse of the same passwords and clicking on malicious links may lead to such types of attacks. When assaulters get entrance to the honest user account, they can grab the movement and abuse existing data and return inaccurate information to the user account. Where small corrections can lead to big disasters in healthcare.

12.5.4 IDENTITY THEFT

In this type of danger, the attackers use the honest user identity to get access to some exclusive benefits, and access those benefits which are intended for legitimate user, when attackers get those benefits and use them against the original user for example when an attacker robs the identity of the user then attackers can book unwanted appointments and also, he can access some private data and make it public, etc.

TABLE 12.1

Possible Threats on IoMT [105]

Threats	Effects	Solutions	References
Data Loss and Data Leakage	Loss of information authenticity, i.e., personal data modified, corrupted etc.	To deal with we need to use the backup mechanism of the cloud	[15, 35, 36, 43, 53, 62, 93–102]
Insecure API and Interfaces	Improper authentication and authorization to access the data	We need to encrypt the transmission and there should be strong access controls	[16, 20, 38, 44, 45]
Malicious Insider	Compromises policies in the healthcare centres to steal users' sensitive personal data	Use breach notification and agreement reporting	[17, 34, 39, 46, 47]
Account hijacking	Here the confidentiality and integrity users' sensitive personal data will be affected once attacker gains access to user account.	Use secure communication channel and strict password change and username change policies with strong authentication mechanisms	[18, 35, 58–62, 88–92]
Identity Theft	The attackers gain identity of a right user to access the user aids and bear befit of the user.	Use strong muti-factor authentication to verify the legitimate user	[19, 37, 42, 48, 49, 52]

12.6 ATTACKS ON THE IOMT

The attacks on the IoT in 2020 touched 10.8 million. Even with a drop in that of 6 million in 2021 [13] we cannot expect a significant drop as health centres around the globe adopt the IoMT for better services. We have listed a few of the most common attacks on the IoMT in Table 12.2.

Zombie Attack (DoS/DDoS Attack): It is a type of attack where attackers send thousands of request packets to the target system via the Internet. Here the main ambition of the attacker is to devour the assets of the victim. This type of attack may influence the benefits of the IoMT. Here is the attacker's flood number of the Transmission Control Protocol (TCP) a packet with the SYN flag set to its target. The victim makes a three-way handshake with the attacker. Here user time will consume to make the connection with the usage of the spaces in the buffer. This is also called an SYN flood attack. Alternatively, the attacker can send a larger number of User Datagram Protocol (UDP) packets to the non-listening ports of the victim. In this case, the victim sends Internet Control Message Protocol (ICMP) message to the attacker. This causes to consume the time of the victim to send a response packet to the attacker. Also in another way, an attacker sends many ICMP "Echo Request" packets to the victim. Here these packets consist of the special broadcast address of

TABLE 12.2

Possible Attacks on IoMT

Attacks	Attack procedures	Effects	Solutions
Zombie attack (DoS/ DDoS attack) [21, 40, 56, 63–66]	Here there is SYN packet flooding to host.	Availability of the critical personal data are affected.	There should be strong authentication and authorization techniques; also there should be strong network monitoring mechanisms.
Man-in-Middle (MIM) attack [22–24, 33, 81–87]	Intervening the conversation between two parties	It influences the confidentiality, integrity and availability of the data in conversion.	Providing proper Secure Socket Layer (SSL) architecture.
Phishing attack [25, 26, 51, 67–72]	By malicious weblinks and redirecting user to fake web pages to collect sensitive information.	It affects the confidentiality of data and also the privacy of the legitimate user	Use the secure web links that uses Hypertext Transfer Protocol Secure (HTTPS) extension.
Root access attack [27, 32, 50, 73–75]	Here whole resources of valid user is exploited.	It affects the sensitive data and services of the user.	Using strong password mechanisms.
Services injection attacks [28, 41, 54, 55]	The legitimate resources are corrupted by malicious code.	Here the valid services provided to the user affected.	There should be strong isolations mechanisms.
Backdoor channel attack [29–31, 76–80]	Using malicious programme to connect remotely with user.	It affects service integrity.	Required strong authentication mechanisms.

the network. Now, the victim sends "Echo reply" messages from every host on its network. This is famously regarded as Smurf Attack, whereas in Distributed Denial-of-Service (DDoS) attackers will scan entire network and identify weak hosts and handlers. These handlers are used to generate or occupy representatives which are called zombies to project the DDoS. [104].

Man-in-Middle (MIM) Attack: In this attack, the assaulter is active in the middle of the association between two parties and accesses the data that are exchanged between them. This is a direct menace to the CIA of the two entities in the conversation as the assailant can view the information, change the data and discard the data. Most often the attack is possible with weak security configuration in Secure Socket Layer (SSL) [104].

Phishing Attacks: The main aim of this is to rob the credentials from the honest user if the benefit from which the assailant gains access to the account can impact the CIA matrix of the data in the user account. In 2021, 27% of organizations around the world detected four to six successful phishing attacks [14]. The attacker sends malicious and manipulated web links called phishing links to the victim. When the victim clicks on the phishing link they are redirected to a fake web page which looks like a legitimate web page. The victim enters his credentials and these are captured by the malicious actor. This type of attack can be avoided by cross-verifying web addresses, avoiding clicking on weblinks from unknown sources and verifying the SSL certificate.

Root Access Attack: In this type of attack, the attacker acquires unlimited access to the entire system. Now the attacker can manipulate the system resources and data in it. This is a direct threat to the user's sensitive information and fails the integrity and availability of the services provided to the users. This can be avoided by adopting strong password mechanisms [104].

Services Injection Attacks: This kind of attack is when the user requests the resource from the cloud. The cloud service is responsible for allocating the requested resources to the legitimate user. The cloud checks for the availability of the resources at a point in time and after the use of the user the same resources are allocated to some other user. Here attacker tries to inject malicious resources into the resources that are allocated by the cloud. Now when a victim requests the resource the cloud serves altered or malicious resources to the victim. This affects the integrity of the resource. This type of attack can be prevented by enforcing an integrity module [104].

Backdoor Channel Attack: Here attacker uses the remote-control programmes that will allow the attacker to alter the resources and data of the victim. This is a direct threat to the confidentiality of the resources. This programme has the potential to use victim resources for DDoS Attacks [104].

12.7 FUTURE SCOPE

The future scope of this IoMT ecosystem is by using deep learning and machine learning methods to classify cardiac seizures and stroke prediction in real time; implementation of IoT-enabled AR and VR devices for treatment in rehabilitation centres; categorization of electronic health records based on readily available data; determining the tremor symptoms by analysing the hand moments; development of new algorithms for Internet of Nano things (IoNT) which makes using nano sensors possible in health care. For hands-on training of medical applications there should be development in signal-processing algorithms to generate feedback based on the input. Communication standards for reliable control of the nano robots should be improved.

12.8 CONCLUSION

The IoMT will be the most prominent field where we can solve many real-world problems in the healthcare sector and also with proper utilization of data, we can reduce many causalities. So, this IoMT generates a lot of critical personal health data. Here we are trying to provide a secure environment for this data and help to secure the storage of this data. The intent is to create an environment for the end-to-end transaction between the patient and his health professional and this can be accessed anywhere, anytime without losing integrity, confidentiality, availability, etc. By all this, we can attend a secure transaction in the healthcare industry enabling the power of the cloud to reduce pressure in healthcare.

REFERENCES

1. "Internet of Things (IoT): Number of Connected Devices Worldwide from 2012 to 2020 (in Billions)," https://www.statista.com/statistics/471264/iot-number-of-connected-devices-worldwide/.
2. "The Complexities of Physician Supply and Demand: Projections from 2019 to 2034, AAMC," June 2021; G. Berlin, M. Lapointe, M. Murphy and J. Wexler, "Assessing the Lingering Impact of COVID-19 on the Nursing Workforce," McKinsey, May 11 2022.
3. M. R. A. Robel, S. Bharati, P. Podder, M. Raihan-Al-Masud and S. Mandal, "Fault Tolerance in Cloud Computing- an Algorithmic Approach," in: Abraham, A., Panda, M., Pradhan, S., GarciaHernandez, L., Ma, K. (eds.) Innovations in Bio-Inspired Computing and Applications. IBICA 2019. *Advances in Intelligent Systems and Computing*, 1180, Springer, Cham, 2021, doi: 10.1007/978-3-030-49339-4_31.
4. M. Papaioannou, M. Karageorgou, G. Mantas, et al., "A Survey on Security Threats and Countermeasures in Internet of Medical Things (IoMT)," *Transactions on Emerging Telecommunications Technologies*, p. e4049, 2020, doi: 10.1002/ett.4049.
5. S. Rani, S. H. Ahmed, R. Talwar, J. Malhotra and H. Song, "IoMT: A Reliable Cross Layer Protocol for Internet of Multimedia Things," *IEEE Internet of Things Journal*, 4(3), pp. 832–839, June 2017, doi: 10.1109/JIOT.2017.2671460.
6. N. Garg, M. Wazid, A. K. Das, D. P. Singh, J. J. P. C. Rodrigues and Y. Park, "BAKMP-IoMT: Design of Blockchain Enabled Authenticated Key Management Protocol for Internet of Medical Things Deployment," *IEEE Access*, 8, pp. 95956–95977, 2020, doi: 10.1109/ACCESS.2020.2995917.
7. "Global Remote Patient Monitoring Devices Market Are Projected to Reach USD 2114 Million at a CAGR of 7.45%, by 2028," https://www.globenewswire.com/en/news-release/2022/11/08/2551121/0/en/Global-Remote-Patient-Monitoring-Devices-Market-Are- Projected-to-Reach-USD-2114-million-at-a-CAGR-of-7-45-by-2028.html.
8. C. Jackson, M. Shahsahebi, T. Wedlake and C. A. DuBard, "Timeliness of Outpatient Follow-Up: An Evidence-Based Approach for Planning after Hospital Discharge," *Annals of Family Medicine*, 13(2), pp. 115–122, March 2015, doi: 10.1370/afm.1753. PMID: 25755032. PMCID: PMC4369604.
9. J. Benbassat and M. Taragin, "Hospital Readmissions as a Measure of Quality of Health Care: Advantages and Limitations," *Archives of Internal Medicine*, 160(8), pp. 1074–1081, 2000, doi: 10.1001/archinte.160.8.1074.
10. I. Bermudez, S. Traverso, M. Mellia and M. Munafò, "Exploring the Cloud from Passive Measurements: The Amazon AWS Case," *2013 Proceedings of the IEEE Infocom*, Turin, pp. 230–234, 2013, doi: 10.1109/INFCOM.2013.6566769.
11. M. Copeland, J. Soh, A. Puca, M. Manning and D. Gollob, "Microsoft Azure and Cloud Computing," in: *Microsoft Azure*, Apress, Berkeley, CA, 2015, doi: 10.1007/978-1-4842-1043-7_1.

12. E. Bisong, "An Overview of Google Cloud Platform Services," in: *Building Machine Learning and Deep Learning Models on Google Cloud Platform*, Apress, Berkeley, CA, 2019, doi: 10.1007/978-1-4842-4470-8_2.

13. "Number of Internet of Things (IoT) Malware Attacks Worldwide from 2020 to 2021, by Month," https://www.statista.com/statistics/1322216/worldwide-internet-of-things-attacks/.

14. "Volume of Successful Phishing Attacks on Businesses Worldwide in 2021," https://www.statista.com/statistics/1149241/share-organizations-worldwide-phishing-attack.

15. R. Singh, "Data Leakage and Security on Cloud Computing," 2022.

16. M. Asghar and A. Amjad, "Securing Insecure Web API's in Cloud Computing," *Mitteilungen Klosterneuburg*, pp. 78–79, 2018.

17. A. J. Duncan, S. Creese and M. Goldsmith, "Insider Attacks in Cloud Computing," *2012 IEEE 11th International Conference on Trust, Security and Privacy in Computing and Communications*, Liverpool, pp. 857–862, 2012, doi: 10.1109/TrustCom.2012.188.

18. C. Calmus, "Account or Service Hijacking in Cloud Computing," 2016.

19. M. Suguna, R. Anusia, S. M. Shalinie and S. Deepti, "Secure Identity Management in Mobile Cloud Computing," *2017 International Conference on Nextgen Electronic Technologies: Silicon to Software (ICNETS2)*, Chennai, pp. 42–45, 2017, doi: 10.1109/ICNETS2.2017.8067894.

20. A. Sirisha and G. G. Kumari, "API Access Control in Cloud Using the Role Based Access Control Model," *Trendz in Information Sciences & Computing (TISC2010)*, Chennai, pp. 135–137, 2010, doi: 10.1109/TISC.2010.5714624.

21. M. Masdari and M. Jalali, "A Survey and Taxonomy of DoS Attacks in Cloud Computing: DoS Attacks in Cloud Computing," *Security and Communication Networks*, 9(16), 2016, doi: 10.1002/sec.1539.

22. G. R. Andreica, L. Bozga, D. Zinca and V. Dobrota, "Denial of Service and Man-in-the-Middle Attacks Against IoT Devices in a GPS-Based Monitoring Software for Intelligent Transportation Systems," *2020 19th RoEduNet Conference: Networking in Education and Research (RoEduNet)*, Bucharest, pp. 1–4, 2020, doi: 10.1109/RoEduNet51892.2020.9324865.

23. J. Choi, B. Ahn, G. Bere, S. Ahmad, H. A. Mantooth and T. Kim, "Blockchain-Based Man-in-the-Middle (MITM) Attack Detection for Photovoltaic Systems," *IEEE Design Methodologies Conference (DMC)*, Bath, pp. 1–6, 2021, doi: 10.1109/DMC51747.2021.9529949.

24. C. -Y. Cheng, E. Colbert and H. Liu, "Experimental Study on the Detectability of Man-in-the-Middle Attacks for Cloud Applications," *2019 IEEE Cloud Summit*, Washington, DC, pp. 52–57, 2019, doi: 10.1109/CloudSummit47114.2019.00015.

25. A. Bhardwaj, S. S. Chandok, A. Bagnawar, S. Mishra and D. Uplaonkar, "Detection of Cyber Attacks: XSS, SQLI, Phishing Attacks and Detecting Intrusion Using Machine Learning Algorithms," *2022 IEEE Global Conference on Computing, Power and Communication Technologies (GlobConPT)*, New Delhi, pp. 1–6, 2022, doi: 10.1109/GlobConPT57482.2022.9938367.

26. M. A. Ivanov, B. V. Kliuchnikova, I. V. Chugunkov and A. M. Plaksina, "Phishing Attacks and Protection Against Them," *2021 IEEE Conference of Russian Young Researchers in Electrical and Electronic Engineering (ElConRus)*, St. Petersburg, pp. 425–428, 2021, doi: 10.1109/ElConRus51938.2021.9396693.

27. A. A. Shaikh, "Attacks on Cloud Computing and Its Countermeasures," *2016 International Conference on Signal Processing, Communication, Power and Embedded System (SCOPES)*, Paralakhemundi, pp. 748–752, 2016, doi: 10.1109/SCOPES.2016.7955539.

28. E. Ijcsis, N. Singh, "SQL Injection Attack Detection & Prevention over Cloud Services," 2016, doi: 10.6084/M9.FIGSHARE.3362392.V1.

29. J. Yang, J. Zheng, H. Wang, J. Li, H. Sun, W. Han, N. Jiang and T. Yu-an, "Edge-Cloud Collaborative Defense Against Backdoor Attacks in Federated Learning," *Sensors*, 23, p. 1052, 2023. doi: 10.3390/s23031052.

30. C. W. Tien, T.-T. Tsai, I.-Y. Chen and S.-Y. Kuo, "UFO - Hidden Backdoor Discovery and Security Verification in IoT Device Firmware," pp. 18–23, 2018, doi: 10.1109/ISSREW.2018.00-37.

31. S. Benzarti, B. Triki and O. Korbaa, "A Survey on Attacks in Internet of Things Based Networks," 2017, doi: 10.1109/ICEMIS.2017.8273006.

32. J. Jiang, Y. Liu and B. Dezfouli, "A Root-Based Defense Mechanism Against RPL Blackhole Attacks in Internet of Things Networks," *2018 Asia-Pacific Signal and Information Processing Association Annual Summit and Conference (APSIPA ASC)*, Honolulu, HI, pp. 1194–1199, 2018, doi: 10.23919/APSIPA.2018.8659504.

33. A. A. Olazabal, J. Kaur and A. Yeboah-Ofori, "Deploying Man-in-the-Middle Attack on IoT Devices Connected to Long Range Wide Area Networks (LoRaWAN)," *2022 IEEE International Smart Cities Conference (ISC2)*, Pafos, Cyprus, pp. 1–7, 2022, doi: 10.1109/ISC255366.2022.9922377.

34. S. S. Tirumala, H. Sathu and V. Naidu, "Analysis and Prevention of Account Hijacking Based INCIDENTS in Cloud Environment," *2015 International Conference on Information Technology (ICIT)*, Bhubaneswar, pp. 124–129, 2015, doi: 10.1109/ICIT.2015.29.

35. K. Kaur, I. Gupta and A. Singh, "A Comparative Evaluation of Data Leakage/Loss Prevention Systems (DLPS)," pp. 87–95, 2017, doi: 10.5121/csit.2017.71008.

36. B. Hauer, "Data and Information Leakage Prevention Within the Scope of Information Security," *IEEE Access*, 3, pp. 1–1, 2015, doi: 10.1109/ACCESS.2015.2506185.

37. K. Turville, J. Yearwood and C. Miller, "Understanding Victims of Identity Theft: Preliminary Insights," *2010 Second Cybercrime and Trustworthy Computing Workshop*, Ballarat, VIC, pp. 60–68, 2010, doi: 10.1109/CTC.2010.12.

38. R. Sun, Q. Wang and L. Guo, "Research Towards Key Issues of API Security," In: Lu, W., Zhang, Y., Wen, W., Yan, H., Li, C. (eds.) *Cyber Security*. CNCERT 2021, *Communications in Computer and Information Science*, 1506, Springer, Singapore, 2022, doi: 10.1007/978-981-16-9229-1_11.

39. M. Omar, "Insider Threats: Detecting and Controlling Malicious Insiders," 2015, doi: 10.4018/978-1-4666-8345-7.ch009.

40. P. Agbedemnab, A.-M. Salifu and Z. Abdulrahim, "Identifying and Isolating Zombie Attack in Cloud Computing," *Asian Journal of Research in Computer Science*, pp. 46–56, 2020, doi: 10.9734/ajrcos/2020/v6i230157.

41. U. Lakhina, "SQL Injection Attack Detection & Prevention over Cloud Services," 2016.

42. J. Winn and K. H. Govern, "Identity Theft: Risks and Challenges to Business of Data Compromise," *Temple Journal of Science, Technology, and Environmental Law*, 28(49), 2009, Available at SSRN: https://ssrn.com/abstract=2093493.

43. S. Nayak and A. Ojha, "Data Leakage Detection and Prevention: Review and Research Directions," 2020, doi: 10.1007/978-981-15-1884-3_19.

44. T. Alladi, V. Chamola, B. Sikdar and K. -K. R. Choo, "Consumer IoT: Security Vulnerability Case Studies and Solutions," *IEEE Consumer Electronics Magazine*, 9(2), pp. 17–25, 1 March 2020, doi: 10.1109/MCE.2019.2953740.

45. X. Wang et al., "Attack and Defence of Ethereum Remote APIs," *2018 IEEE Globecom Workshops (GC Wkshps), Abu Dhabi*, pp. 1–6, 2018, doi: 10.1109/GLOCOMW.2018.8644498.

46. J. Glasser and B. Lindauer, "Bridging the Gap: A Pragmatic Approach to Generating Insider Threat Data," *2013 IEEE Security and Privacy Workshops, San Francisco, CA*, pp. 98–104, 2013, doi: 10.1109/SPW.2013.37.

47. K. Nance and R. Marty, "Identifying and Visualizing the Malicious Insider Threat Using Bipartite Graphs," *2011 44th Hawaii International Conference on System Sciences*, Kauai, HI, pp. 1–9, 2011, doi: 10.1109/HICSS.2011.231.

48. Y. Cho and S. Lee, "Detection and Response of Identity Theft within a Company Utilizing Location Information," *2016 International Conference on Platform Technology and Service (PlatCon)*, Jeju, Korea (South), pp. 1–5, 2016, doi: 10.1109/PlatCon.2016.7456790.

49. G. Kolaczek, "An Approach to Identity Theft Detection Using Social Network Analysis," *2009 First Asian Conference on Intelligent Information and Database Systems*, Dong Hoi, pp. 78–81, 2009, doi: 10.1109/ACIIDS.2009.44.

50. Z. Chen, F. Han, J. Cao, X. Jiang and S. Chen, "Cloud Computing-Based Forensic Analysis for Collaborative Network Security Management System," *Tsinghua Science and Technology*, 18(1), pp. 40–50, February 2013, doi: 10.1109/TST.2013.6449406.

51. J. Dawkins and J. Hale, "A Systematic Approach to Multi-stage Network Attack Analysis," *Second IEEE International Information Assurance Workshop, 2004. Proceedings, Charlotte, NC*, pp. 48–56, 2004, doi: 10.1109/IWIA.2004.1288037.

52. S. G. A. van de Weijer, R. Leukfeldt and W. Bernasco, "Determinants of Reporting Cybercrime: A Comparison between Identity Theft, Consumer Fraud, and Hacking," *European Journal of Criminology*, 16(4), pp. 486–508, 2019, doi: 10.1177/1477370818773610.

53. A. A. Christina, "Proactive Measures on Account Hijacking in Cloud Computing Network," *Asian Journal of Computer Science and Technology*, 4(2), pp. 31–34, 2015.

54. N. Gruschka and M. Jensen, "Attack Surfaces: A Taxonomy for Attacks on Cloud Services," *2010 IEEE 3rd International Conference on Cloud Computing*, Miami, FL, pp. 276–279, 2010, doi: 10.1109/CLOUD.2010.23.

55. K. Wang and Y. Hou, "Detection Method of SQL Injection Attack in Cloud Computing Environment," *2016 IEEE Advanced Information Management, Communicates, Electronic and Automation Control Conference (IMCEC)*, Xi'an, pp. 487–493, 2016, doi: 10.1109/IMCEC.2016.7867260.

56. A. Chonka and J. Abawajy, "Detecting and Mitigating HX-DoS Attacks Against Cloud Web Services," *2012 15th International Conference on Network-Based Information Systems*, Melbourne, VIC, pp. 429–434, 2012, doi: 10.1109/NBiS.2012.146.

57. S. Tarikere, I. Donner and D. Woods, "Diagnosing a Healthcare Cybersecurity Crisis: The Impact of IoMT Advancements and 5G," *Business Horizons*, 64(6), pp. 799–807, 2021, ISSN:0007 6813, doi: 10.1016/j.bushor.2021.07.015. (https://www.sciencedirect.com/science/article/pii/S0007681321001385).

58. C. Prakash and S. Dasgupta, "Cloud Computing Security Analysis: Challenges and Possible Solutions," *2016 International Conference on Electrical, Electronics, and Optimization Techniques (ICEEOT)*, Chennai, pp. 54–57, 2016, doi: 10.1109/ICEEOT.2016.7755626.

59. N. C. Paxton, "Cloud Security: A Review of Current Issues and Proposed Solutions," *2016 IEEE 2nd International Conference on Collaboration and Internet Computing (CIC)*, Pittsburgh, PA, pp. 452–455, 2016, doi: 10.1109/CIC.2016.066.

60. L. Alhenaki, A. Alwatban, B. Alamri and N. Alarifi, "A Survey on the Security of Cloud Computing," *2019 2nd International Conference on Computer Applications & Information Security (ICCAIS)*, Riyadh, Saudi Arabia, pp. 1–7, 2019, doi: 10.1109/CAIS.2019.8769497.

61. R. A. Nafea and M. Amin Almaiah, "Cyber Security Threats in Cloud: Literature Review," *2021 International Conference on Information Technology (ICIT)*, Amman, pp. 779–786, 2021, doi: 10.1109/ICIT52682.2021.9491638.

62. M. Derfouf, A. Mimouni and M. Eleuldj, "Vulnerabilities and Storage Security in Cloud Computing," *2015 International Conference on Cloud Technologies and Applications (CloudTech)*, Marrakech, pp. 1–5, 2015, doi: 10.1109/CloudTech.2015.7337002.

63. M. Darwish, A. Ouda and L. F. Capretz, "Cloud-Based DDoS Attacks and Defenses," *International Conference on Information Society (i-Society 2013)*, Toronto, ON, pp. 67–71, 2013.

64. S. Potluri, M. Mangla, S. Satpathy and S. N. Mohanty, "Detection and Prevention Mechanisms for DDoS Attack in Cloud Computing Environment," *2020 11th International Conference on Computing, Communication and Networking Technologies (ICCCNT)*, Kharagpur, pp. 1–6, 2020, doi: 10.1109/ICCCNT49239.2020.9225396.

65. B. Prabadevi and N. Jeyanthi, "Distributed Denial of Service Attacks and Its Effects on Cloud Environment- a Survey," *The 2014 International Symposium on Networks, Computers and Communications*, Hammamet, pp. 1–5, 2014, doi: 10.1109/ SNCC.2014.6866508.

66. G. Somani, M. S. Gaur, D. Sanghi, M. Conti, M. Rajarajan and R. Buyya, "Combating DDoS Attacks in the Cloud: Requirements, Trends, and Future Directions," *IEEE Cloud Computing*, 4(1), pp. 22–32, January–February 2017, doi: 10.1109/MCC.2017.14.

67. N. Agrawal and S. Tapaswi, "Defense Mechanisms Against DDoS Attacks in a Cloud Computing Environment: State-of-the-Art and Research Challenges," *IEEE Communications Surveys and Tutorials*, 21(4), pp. 3769–3795, Fourthquarter 2019, doi: 10.1109/COMST.2019.2934468.

68. R. Alabdan, "Phishing Attacks Survey: Types, Vectors, and Technical Approaches," *Future Internet*, 12(10), p. 168, 2020, doi: 10.3390/fi12100168.

69. M. T. Khorshed, A. B. M. S. Ali and S. A. Wasimi, "Trust Issues That Create Threats for Cyber Attacks in Cloud Computing," *2011 IEEE 17th International Conference on Parallel and Distributed Systems*, Tainan, pp. 900–905, 2011, doi: 10.1109/ ICPADS.2011.156.

70. A. O'Mara, I. Alsmadi and A. AlEroud, "Generative Adverserial Analysis of Phishing Attacks on Static and Dynamic Content of Webpages," *2021 IEEE International Conference on Parallel & Distributed Processing with Applications, Big Data & Cloud Computing, Sustainable Computing & Communications, Social Computing & Networking (ISPA/BDCloud/SocialCom/SustainCom)*, New York, pp. 1657–1662, 2021, doi: 10.1109/ISPA-BDCloud-SocialCom-SustainCom52081.2021.00222.

71. M. Jensen, J. Schwenk, N. Gruschka and L. L. Iacono, "On Technical Security Issues in Cloud Computing," *2009 IEEE International Conference on Cloud Computing*, Bangalore, pp. 109–116, 2009, doi: 10.1109/CLOUD.2009.60.

72. W. Yao, Y. Ding and X. Li, "Deep Learning for Phishing Detection," *2018 IEEE International Conference on Parallel & Distributed Processing with Applications, Ubiquitous Computing & Communications, Big Data & Cloud Computing, Social Computing & Networking, Sustainable Computing & Communications (ISPA/IUCC/ BDCloud/SocialCom/SustainCom)*, Melbourne, VIC, pp. 645–650, 2018, doi: 10.1109/ BDCloud.2018.00099.

73. S. Mandal and D. A. Khan, "A Study of Security Threats in Cloud: Passive Impact of COVID-19 Pandemic," *2020 International Conference on Smart Electronics and Communication (ICOSEC)*, Trichy, pp. 837–842, 2020, doi: 10.1109/ ICOSEC49089.2020.9215374.

74. T. Combe, A. Martin and R. Di Pietro, "To Docker or Not to Docker: A Security Perspective," *IEEE Cloud Computing*, 3(5), pp. 54–62, September–October 2016, doi: 10.1109/MCC.2016.100.

75. M. S. Ayyub and M. P. Kaushik, "An Analysis of Security Attacks on Cloud Wrt SaaS," *International Journal of Advancements in Research & Technology*, 4(2), pp.81–86, 2015.

76. E. N. Saad, K. E. Mahdi and M. Zbakh, "Cloud Computing Architectures Based IDS," *2012 IEEE International Conference on Complex Systems (ICCS)*, Agadir, pp. 1–6, 2012, doi: 10.1109/ICoCS.2012.6458581.

77. M. Durairaj and A. Manimaran, "A Study on Security Issues in Cloud Based E-learning," *Indian Journal of Science and Technology*, 8(8), pp. 757–765, 2015.

78. A. Patel, N. Shah, D. Ramoliya and A. Nayak, "A Detailed Review of Cloud Security: Issues, Threats & Attacks," *2020 4th International Conference on Electronics, Communication and Aerospace Technology (ICECA)*, Coimbatore, pp. 758–764, 2020, doi: 10.1109/ICECA49313.2020.9297572.

79. C. Xu, W. Liu, Y. Zheng, S. Wang and C. -H. Chang, "Inconspicuous Data Augmentation Based Backdoor Attack on Deep Neural Networks," *2022 IEEE 35th International System-on-Chip Conference (SOCC)*, Belfast, pp. 1–6, 2022, doi: 10.1109/SOCC56010.2022.9908113.

80. Y. Luo, W. Luo, X. Sun, Q. Shen, A. Ruan and Z. Wu, "Whispers between the Containers: High-Capacity Covert Channel Attacks in Docker," *2016 IEEE Trustcom/ BigDataSE/ISPA*, Tianjin, pp. 630–637, 2016, doi: 10.1109/TrustCom.2016.0119.

81. M. T. S. R. Jat and S. Sharma, "Survey on Cloud Attack Types and Detection Techniques," *International Journal of Scientific Research & Engineering Trends*, 5, p. 6, 2021.

82. A. Rauf, R. A. Shaikh and A. Shah, "Security and Privacy for IoT and Fog Computing Paradigm," *2018 15th Learning and Technology Conference (L&T)*, Jeddah, pp. 96–101, 2018, doi: 10.1109/LT.2018.8368491.

83. G. Rathee, A. Sharma, R. Kumar, F. Ahmad and R. Iqbal, "A Trust Management Scheme to Secure Mobile Information Centric Networks," *Computer Communications*, 151, pp. 66–75, 2020, ISSN01403664, doi: 10.1016/j.comcom.2019.12.024. (https:// www.sciencedirect.com/science/article/pii/S0140366419309934).

84. I. O. Ogundele, A. O. Akinade, H. O. Alakiri, A. A. Aromolaran and B. O. Uzoma, "Detection and Prevention of Session Hijacking in Web Application Management," *The International Journal of Advanced Research in Computer and Communication Engineering*, 9(6), pp. 1–10, 2020.

85. M. Denis, C. Zena and T. Hayajneh, "Penetration Testing: Concepts, Attack Methods, and Defense Strategies," *2016 IEEE Long Island Systems, Applications and Technology Conference (LISAT)*, Farmingdale, NY, pp. 1–6, 2016, doi: 10.1109/ LISAT.2016.7494156.

86. S. Itoo, A. A. Khan, V. Kumar, A. Alkhayyat, M. Ahmad and J. Srinivas, "CKMIB: Construction of Key Agreement Protocol for Cloud Medical Infrastructure Using Blockchain," *IEEE Access*, 10, pp. 67787–67801, 2022, doi: 10.1109/ ACCESS.2022.3185016.

87. T. Akhtar and B. B. Gupta, "Towards a Framework for Analyzing Cyber Attacks Impact Against Smart Power Grid on SCADA System," *2018 International Conference on Communication and Signal Processing (ICCSP)*, Chennai, pp. 1087–1093, 2018, doi: 10.1109/ICCSP.2018.8524195.

88. A. A. Nayak, N. K. Sridhar, G. R. Poornima and Shivashankar, "Security Issues in Cloud Computing and Its Counter Measure," *2017 2nd IEEE International Conference on Recent Trends in Electronics, Information & Communication Technology (RTEICT)*, Bangalore, pp. 35–41, 2017, doi: 10.1109/RTEICT.2017.8256554.

89. W. Alnahari and M. T. Quasim, "Authentication of IoT Device and IoT Server Using Security Key," *2021 International Congress of Advanced Technology and Engineering (ICOTEN)*, Taiz, pp. 1–9, 2021, doi: 10.1109/ICOTEN52080.2021.9493492.

90. Y. Shah and S. Sengupta, "A Survey on Classification of Cyber-Attacks on IoT and IIoT Devices," *2020 11th IEEE Annual Ubiquitous Computing, Electronics & Mobile Communication Conference (UEMCON)*, New York, pp. 0406–0413, 2020, doi: 10.1109/UEMCON51285.2020.9298138.

91. W. Yasin and N. Jayapandian, "A Review on Cyber Security Issues and Research Challenges in Internet of Things," *2021 4th International Conference on Recent Trends in Computer Science and Technology (ICRTCST)*, Jamshedpur, pp. 348–353, 2021, doi: 10.1109/ICRTCST54752.2022.9782046.

92. T. Choudhury, A. Gupta, S. Pradhan, P. Kumar and Y. S. Rathore, "Privacy and Security of Cloud-Based Internet of Things (IoT)," *2017 3rd International Conference on Computational Intelligence and Networks (CINE)*, Odisha, pp. 40–45, 2017, doi: 10.1109/CINE.2017.28.

93. A. Badhib, S. Alshehri and A. Cherif, "A Robust Device-to-Device Continuous Authentication Protocol for the Internet of Things," *IEEE Access*, 9, pp. 124768–124792, 2021, doi: 10.1109/ACCESS.2021.3110707.

94. Q. B. Hani and J. P. Dichter, "Data Leakage Preventation Using Homomorphic Encryptionin Cloud Computing," *2016 IEEE Long Island Systems, Applications and Technology Conference (LISAT)*, Farmingdale, NY, pp. 1–5, 2016, doi: 10.1109/LISAT.2016.7494153.

95. R. Latif, H. Abbas, S. Assar and Q. Ali, "Cloud Computing Risk Assessment: A Systematic Literature Review," *Future Information Technology*, pp. 285–295, 2014.

96. K. Dahbur, B. Mohammad and A. B. Tarakji, "A Survey of Risks, Threats and Vulnerabilities in Cloud Computing," *Proceedings of the 2011 International Conference on Intelligent Semantic Web-Services and Applications*, pp. 1–6, April 2011.

97. R. Wang, "Research on Data Security Technology Based on Cloud Storage," *Procedia Engineering*, 174, pp. 1340–1355, 2017.

98. P. A. Boampong and L. A. Wahsheh, "Different Facets of Security in the Cloud," *Proceedings of the 15th Communications and Networking Simulation Symposium*, pp. 1–7, March 2012.

99. X. Yu and Q. Wen, "A View about Cloud Data Security from Data Life Cycle," *2010 International Conference on Computational Intelligence and Software Engineering*, Wuhan, pp. 1–4, 2010, doi: 10.1109/CISE.2010.5676895.

100. Y. Yu et al., "Identity-Based Remote Data Integrity Checking With Perfect Data Privacy Preserving for Cloud Storage," *IEEE Transactions on Information Forensics and Security*, 12(4), pp. 767–778, April 2017, doi: 10.1109/TIFS.2016.2615853.

101. M. U. Bokhari, Q. M. Shallal and Y. K. Tamandani, "Cloud Computing Service Models: A Comparative Study," *2016 3rd International Conference on Computing for Sustainable Global Development (INDIACom)*, New Delhi, pp. 890–895, 2016.

102. X. Zhang, C. Liu, S. Nepal, S. Pandey and J. Chen, "A Privacy Leakage Upper Bound Constraint-Based Approach for Cost-Effective Privacy Preserving of Intermediate Data Sets in Cloud," *IEEE Transactions on Parallel and Distributed Systems*, 24(6), pp. 1192–1202, June 2013, doi: 10.1109/TPDS.2012.238.

103. A. U. Khan, M. Oriol, M. Kiran, M. Jiang and K. Djemame, "Security Risks and Their Management in Cloud Computing," *4th IEEE International Conference on Cloud Computing Technology and Science Proceedings*, Taipei, pp. 121–128, 2012, doi: 10.1109/CloudCom.2012.6427574.

104. A. Ghubaish, T. Salman, M. Zolanvari, D. Unal, A. Al-Ali and R. Jain, "Recent Advances in the Internet-of-Medical-Things (IoMT) Systems Security," *IEEE Internet*

of Things Journal, 8(11), pp. 8707–8718, June 1 2021, doi: 10.1109/JIOT.2020. 3045653.

105. A. Singh and K. Chatterjee, "Cloud Security Issues and Challenges: A Survey," *Journal of Network and Computer Applications*, 79, pp. 88–115, 2017, ISSN 1084-8045, doi: 10.1016/j.jnca.2016.11.027. (https://www.sciencedirect.com/science/article/pii/S1084804516302983).

Index

191

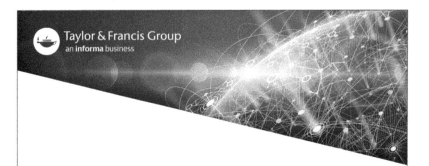

Milton Keynes UK
Ingram Content Group UK Ltd.
UKHW031132141024
449569UK00006B/249